The Tactical Advantage

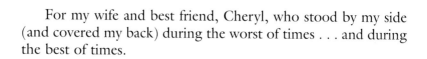

For my wife and best friend, Cheryl, who stood by my side (and covered my back) during the worst of times . . . and during the best of times.

The Tactical Advantage

A DEFINITIVE STUDY OF PERSONAL SMALL-ARMS TACTICS

Gabriel Suarez

PALADIN PRESS • BOULDER, COLORADO

Also by Gabriel Suarez:

Close-Range Gunfighting (video)
Combative Perspective
Force-on-Force Gunfight Training
Force-on-Force Handgun Drills (video)
Tactical Advantage: The Video
Tactical Pistol
Tactical Pistol Marksmanship
Tactical Rifle
Tactical Shotgun

The Tactical Advantage:
A Definitive Study of Personal Small-Arms Tactics
by Gabriel Suarez

Copyright © 1998 by Gabriel Suarez

ISBN 13: 978-0-87364-975-9
Printed in the United States of America

Published by Paladin Press, a division of
Paladin Enterprises, Inc.
Gunbarrel Tech Center
7077 Winchester Circle
Boulder, Colorado 80301 USA
+1.303.443.7250

Direct inquiries and/or orders to the above address.

PALADIN, PALADIN PRESS, and the "horse head" design
are trademarks belonging to Paladin Enterprises and
registered in United States Patent and Trademark Office.

Visit our Web site at www.paladin-press.com

Very Good Book

TABLE OF CONTENTS

FOREWORD

————◆◆◆————

Put your heart and soul into being expert killers with your weapons.
The only good enemy is a dead enemy. Misses do not kill, but a bullet in the
heart or a bayonet in the guts does. Let every bullet find its billet—in the
body of your foes.

General George S. Patton, Jr., December 1941
The Patton Papers, Vol. II, 1974

As a full-time professional weapons and tactics writer, instructor, and consultant, I encounter an amazing number of people who profess knowledge and expertise in a field of endeavor that can only be described as being diverse, abstract, at times subjective, and always dangerous. As you might expect, the written material and instructional programs produced by many of these "experts" reflect an alarming lack of cognizance of real-world concerns, things that can only be known by one who has "been there and done it."

Tactical liability hazards can't be understood by just reading a book or two. Nor can written material dealing with them simply be paraphrased from someone else's articles or, worse, merely theorized upon and then subsequently treated as though it were fact. Yet, the vast majority of the written and instructional data we see these days lacks this critical perspective. As a result, it is usually ineffective—and often downright deadly—if used in a real life-and-death situation.

To make matters worse, a frightening number of the techniques and tactics espoused by these people—those with what I call "the firing range mentality"—were not created with a knowledge of criminal and civil liability issues, thus creating a nightmare of legal implications that can destroy you as completely as a bullet between the eyes.

In the pages that follow, author Gabe Suarez treats the subject of tactics with the respect it not only deserves, but demands. For Gabe is one of those who has "been there and done it," not just once, but a number of times. A law-enforcement Medal of Valor winner for his actions under fire, he has served with distinction as a police officer for more than 10 years in not only patrol functions, but in gang-enforcement and SWAT assignments as well.

In addition, he is one of the best combat pistol shots in the country and handles a submachine gun or assault rifle with the grace and ease of Babe Ruth hitting a home run. He is also highly articulate, making him one of the most effective instructors I have ever seen. In short, Gabe is a warrior of stature, deservedly recognized, admired, and respected by those around him. In this, his third book, read what he has to say about tactics thoroughly and carefully, because he knows what he is talking about.

It has been said that the true professional takes his work seriously, not himself too seriously, and that there is more to being a professional than taking money for services rendered. Believe it, because it's true. Professionalism is a state of mind, a personal creed requiring honesty, objectivity, courage, skill, and articulateness. Gabe Suarez has all of these attributes and has, in this book, provided a clear and realistic look at one of the most important elements of survival in a life-and-death environment.

Chuck Taylor, 1998
Prescott, Arizona

ACKNOWLEDGMENTS

First of all I must thank a special group of gentlemen and warrior-scholars with varied backgrounds who contributed to the completion of this book in many ways. Specifically by name, "The Master's Master" Chuck Taylor, Don Busse, Marc Fleischmann, Dale Fricke, Greg Nordyke, Bruce Parkin, Naish Piazza, and Dave "Top" Teague. A special thanks to Chuck for his technical, tactical, and philosophical advice, as well as for the photographic assistance.

Of course, special thanks must go also to my wife and best friend—the tactical photographer, *pistolera*, and proofreader extraordinaire—for her immense help and patience in all aspects of this text.

Special mention to my good friends Dean Gamburd and Jeff Mayberry for their help with the photographs.

Finally, to my brothers-in-arms—those operators whom I've stood beside, sweated and bled with, gone through the door with, and seen the elephant with. You know who you are, and you will never be forgotten.

INTRODUCTION

A mysterious fraternity born out of smoke and danger of death.
Stephen Crane, 1893
The Red Badge of Courage

As the typical student of small arms progresses in his studies, he eventually realizes that winning a fight requires much more than just good shooting or dexterous gunhandling. He realizes that even an ace tactician will die if he puts himself into an no-win situation. He also begins to realize that how he handles a tactical problem has as much effect on his victory (or defeat) as does the quality of his shooting.

The study of avoiding and, if necessary, managing these situations is what makes up the art of tactics. Notice that I did not say the science of tactics. Tactics are not an exact science like mathematics or physics. There is no "if A does this, then B does that" thinking in this art. The art of tactics is a living one that is always changing and adapting to the needs of the situation. It is dependent for success only on the operator's creativity. Just as there are no guarantees in a gunfight, there are also no absolutes to tactics (you will see this again in Chapter 8).

The study of tactics begins with the study of certain principles and guidelines. I must point out, however, that these principles and guidelines do not guarantee success. Instead, they serve only to help minimize your exposure to danger when ol' Murphy is feeling mischievous.

Some of these concepts were learned from Chuck Taylor at the various courses we taught together. Others, however, were learned through trial and error in the arena where losers die and the only trophy is the right to keep breathing. These concepts were developed from the deadly life-ending mistakes of many of my police comrades . . . and some, from living through many of my own "lucky" mistakes. These were all extremely poignant and valuable lessons learned in the world's toughest school where much of the tuition was paid for with the blood of heroes. In studying these concepts, we attempt to ensure that the next "mishap" happens only to the bad guys, instead of to one of us!

Realize also that the concepts described for dealing with a particular tactical problem (for example, a corner) also apply to any other obstacle that creates the same tactical effect, such as furniture or a bend in a trail.

The result is, in my opinion, the most complete study of tactical principles ever written. But you must be the final judge of that when the last word has been read.

This book is for everyone interested in the subject of personal security, regardless of occupation. It will be of value to the lone homeowner who wants to find the best way to check out that frightening and suspicious noise from across a dark house on a warm summer night. It is for the police officer and his partner who will be searching a building together, and alone, in response to a silent burglary alarm, at 0'dark thirty. It is also for the tactical operative, weighed down by his assault gear, who will be conducting a covert movement to contact with his team during an early-morning raid. I have been in all those pairs of boots many times and know well the needs of each situation.

I will discuss the principles and guidelines that facilitate tactical problem solving, as well as the various architectural features found in many urban dwellings and the best ways to "clear" them, whether you are alone or accompanied by one or two other shooters. I will examine the low-light environment and how to use the available ambient and artificial light to your advantage. Other topics include tactical communications, mindset, the handling of captured hostiles, and the weapons most suitable for the task of searching a dwelling.

Sound tactics are often the only thing that stands between a glorious and spectacular victory and the cold anonymity of death and defeat. Study these principles, incorporate them into your tactical "bag of tricks," and be prepared to win!

Gabriel Suarez, 1998
Santa Monica, California

ONE

TACTICAL MIND-SET

Also I heard the voice of the Lord, saying, "Whom shall I send? And who will go for us?" Then said I, "Here I am. Send me."

Isaiah 6:8

Every student of weaponcraft who has followed the discipline of arms for any length of time knows that shooting skill alone is not enough. The mental attitude, or mind-set, you have when you are faced with violence has as much, and probably more, effect on the outcome as any other factor. Those of you who've read my other books know that I am a vociferous proponent of an aggressive state of mind when the "flag flies."

The only problem is that the flag doesn't fly every day, does it? Even a man who lives in a very hostile environment rarely faces more than two or three deadly confrontations in his life. The problem really becomes one of situational awareness, of actually knowing when a fight is either in progress or about to start. A properly conditioned mind and a fighting attitude will go far in this department.

The first step is the realization that everybody in the world is not a polite, hard-working patriot like you. There are predators out there in the concrete jungle, just as in any jungle. These predators have no regard for your life, your family, or your possessions. They will kill you for the Rolex watch on your arm with as much afterthought as they'd give to eating a cheeseburger. They will cut off your child's finger just to steal her ring. They will not show you or your family any mercy, so do not expect any . . . or give any.

The mental attitude you have when faced with violence has more effect on the outcome of the incident than anything else. (Photo courtesy of Chuck Taylor.)

The second step in the mental development of a nonvictim is the willingness to defend oneself when offered violence. An old gunfighter once said that it wasn't enough to be accurate or fast, one had to be willing. Simply stated, you must be willing to kill any man who would harm you or your family. You must be willing to offer greater violence for violence offered. That attitude must precede all else, so develop it.

Fine, so we know that there are very bad people out there, and we are willing to deck them flat when they confront us. Also, to facilitate our combative responses, we've developed our martial abilities to match our willingness. The question remains, however, how will we know when to act? We simply cannot go through life poised to draw and shoot everyone who we suspect *might* be a predator. We cannot anticipate specifically, so we must anticipate generally. We must develop a state of mind founded on environmental awareness where the sudden appearance of a hostile adversary doesn't surprise us. When confronting an assailant, our thought should not be, "Oh my God, is that awful man really breaking into my house? Is that a real gun in his hand?" Instead it should be, "I see him breaking into my house, he has a gun, and I'm ready for him. Boy, has he made a big mistake!" Attitude makes all the difference.

We must develop a state of mind in which the sudden appearance of a threat is not unexpected.

Most 20th-century human beings, however, are extremely reluctant to harm another person, even when that person has taken clearly overt hostile actions toward them. There are various cultural reasons for this phenomenon, one of which is simply the disbelief that another person would really want to hurt us. Such a mind-set must be overcome if we want to live to tell about it when we have to shoot for our lives.

To do this, you must develop an escalating state of alertness and, subsequently, an awareness of the environment in which you operate. This will provide you with the ability to respond without hesitation when you are confronted with violence. Additionally, such a heightened state of environmental awareness will prevent some of the tragic overreactions that the media like to showcase as reasons for unilateral public disarmament.

The best method I know to obtain such a combative and aware mind-set is through intense study of the color code of readiness. During the research for this book, I learned that the concept of the color code of readiness dates back to the 82nd Airborne Division during World War II. The code was subsequently modified and applied to the realm of personal combat by Jeff Cooper some years after the war.

The first mental state is simply unreadiness. In this condition, all your focus is within yourself on your private thoughts and problems, and you are completely oblivious to your surroundings. This mental state is characterized by the color white (it was originally characterized by the color green). Criminals love to come across people in condition white . . . they make such easy prey.

The next ascending level of alertness is characterized by the color yellow. A man in condition yellow is mentally relaxed, but he is aware of his surroundings. He knows what is behind him, as well as anything that appears unusual or out of place. He watches people as he moves throughout his day, whereas a man in condition white will often have his eyes on the deck. Almost all fights are preceded by subtle clues that a man in condition white will miss. A man in condition yellow will always notice them, because he is paying attention. Chances are good that these clues may be harmless, but he is on guard in the event that they are a prelude to combat. A man in condition yellow realizes that he might have to fight to defend his life, but he doesn't know when this will happen or who his enemy will be. The main difference between a man in condition white and a man in condition yellow is that the latter is paying attention to his immediate surroundings.

After condition yellow, we come to a condition of specific alert, characterized by the color orange. A man in condition orange has noticed one of those prefight clues and is specifically alert to its source. He realizes not only that he may have to shoot, but that he may now have a specific target. Condition orange brings us one step closer to the shooting decision. It is relatively easy to shift mental gears from yellow to orange, but not from white to orange. In terms of conducting a tactical building search, we hunt in yellow (general alert) and go to orange (specific alert) when we close in on specific danger areas. We may not have an actual human target, but the time and place of the fight may very well be in the next few seconds, even if we haven't actually seen the enemy yet.

If we locate anyone concealed in that danger area, we move up the ladder of alertness with subsequent ease in the decision to shoot. This final level is condition red, which means that a fight is now quite likely. We haven't decided to shoot yet, but we've located a specific individual who may be hostile and who may require shooting, depending on his reaction to us. The determining factor to your response is the personal establishment of a mental trigger. This is simply your perception of the subject's intent based on his actions. This may include perception of a weapon of any sort in his hands or an aggressive move toward you, or in some extreme cases, an actual shot fired at you. The mental trigger that you establish is limited only by your legal and moral rules of engagement. You must establish your mental trigger long before the fight so that when the event unfolds, you will not require a personal debate about whether you should shoot. Although the response is a conscious decision, it is almost instantaneous, like a conditioned reflex.

When the fight begins, you must pay complete and undivided attention to solving the problem at hand. This means simply using proper tactics and shooting well and requires extreme concentration on the task at hand. You must not dwell on any shots you may have missed or a faulty tactic that you may have used. Neither do you plan ahead to the next shot to be fired. Instead, concentrate on and experience the shot you are firing—right now! You will, of course, have to plan ahead as far as how to deal with a tactical problem, such as a room search, a door entry, or a flanking maneuver, but this does not occur as you are defending yourself. In any event, once the fight begins, you must not hesitate.

Your level of apprehension prior to beginning a tactical problem will also dictate your state of mind during its solution. Be specific with yourself about why you are there and what it is you are trying to accomplish. Are you simply investigating a suspicious noise, or is it an obvious home invasion? Are you conducting a low-risk administrative perimeter check, or are you hunting

Your response to a threat must be "front sight, press!"

for a hidden and armed criminal? Are you dealing with a simple open door that someone forgot to close, or were three armed gang members seen breaking it down seconds earlier? Each of these scenarios is different in its degree of perceived danger. Knowing what you are getting into and what you are trying to accomplish may dictate the tactics you will choose. For example, you may elect to take a shotgun or submachine gun instead of a pistol. You may want to grab a flashlight or call for reinforcements before commencing. You might even want to hide quietly in the darkness and wait for them to come to you where you can deliver the terminal surprise.

During a tactical operation you are in condition yellow. When you locate a target indicator or you commit to clearing a danger area, you go into condition orange. When you contact the source of the target indicator, you are in condition red, and all systems are go.

To be in any other state of mind is . . . well, suicide. The decision to shoot depends on the actions and disposition of the hostile. Your perception of this is the mental trigger. You will now be looking for any action that you've predetermined to be enough of a threat to warrant a lethal response. If the subject takes such action (e.g., gun in his hand or pointed at you), shoot him. End of story.

Rules of engagement will vary with the landscape, but generally a hostile with a gun in his hand will get shot. So will a hostile who rushes out to grab you while you are holding him at gunpoint, even if he is unarmed! Some may argue that this is not fair play. So what? If you want fair, go to a boxing match. The graveyards are full of men who believed in fair play. We cannot concern ourselves with such imaginary foolishness in life-or-death combat. Remember, if you lose, you die!

Study the development of this mind-set. It will work wonders for you when you find yourself hunting goblins . . . or when they are hunting you.

Here's a quick test to gauge the level of your mind-set development and readiness for violence. Right now, this very moment, the Manson gang is breaking in through your back door, machetes in hand and murder in their hearts! Where is your gun? Can you get to it in two seconds? One, one thousand—two, one thousand . . .

If you didn't know where your gun was or you couldn't get to it in time, you are now eternally and horizontally D-E-A-D! The penalty: reread this chapter 50 times each night before bedtime for a week. Sweet dreams, grasshopper!

TWO

BASIC TACTICAL PRINCIPLES

———◆———

He shall dwell on high: his place of defense shall be the munitions of rocks: bread shall be given him; his waters shall be sure.

Isaiah 33:16

One of the deadliest misconceptions in tactical circles is that a single operator can safely negotiate an entire area (indoor or outdoor) alone. This point is easy to illustrate when you realize that you cannot look in two directions at once, and searching alone often requires doing just that. A single operator must often turn his back on one danger area to search another. The rule is to *avoid searching alone*.

A police officer or tactical team member will have the option of calling for reinforcements to help in the search. An individual, such as a homeowner, generally does not have that luxury. If he is certain an invasion has occurred, he is better off to barricade himself in a "safe room" and lie in wait for the invaders to come to him. This way he can deal with them by surprise and from cover. The principles of military operations in urban terrain teach us that built-up areas overwhelmingly favor the defender over the aggressor. Thus, plan A for a private citizen/homeowner is to take a covered position and call the police. The same plan for a single police officer is to take a position of advantage outside and call for a backup or two.

But what if a homeowner is not sure enough to call the police? Reality tells us that a citizen who cries wolf at every little sound in the night will receive increasingly delayed police responses (if any at all). What if he isn't sure whether that sound of breaking glass was the cat knocking over a vase or something more sinister? Let's be honest—you probably won't call when you're not sure. Similarly, neither will you lie in wait with your Benelli Super-90 pointed at the triple-locked door, barricaded until the sun rises. Here's another more sobering consideration: what if that sound of glass breaking came from your child's bedroom on the other side of the house? I would go out into the darkness alone to make sure; unless you are a spineless liberal coward, so would you.

With that decision out of the way, I'll discuss some tactical principles that will help minimize the already considerable amount of danger. Remember the reason you are searching, because that will determine the intensity and method of your efforts. If you are simply investigating the sound of glass breaking at the far end of the house, you won't necessarily be looking into closets or under furniture—not yet anyway. You will be moving carefully and stealthily toward the source of the sound. On the flip side, what if that sound of glass breaking was followed by your child screaming and a stranger's foul language ordering her to shut up? You will be conducting a dynamic hostage rescue and be attacking the problem violently and aggressively. Just as you match the speed of your shooting to the difficulty of the problem at hand, so do you balance the level of aggressiveness and degree of speed to the tactical situation you face. Regardless of the type of search you conduct, these principles will help you immensely.

TACTICAL GUIDELINES

1. *Use your senses to look for target indicators.* Primarily, you will rely on your sight and hearing. Do not, however, dismiss the other senses. Your senses of smell and touch will provide essential data about the whereabouts of the enemy during your search.

Target indicators are anything that will point to the presence of an adversary. They are often categorized as shine, movement, sound, smell, shape, contrast, human sign, and tactical sign. Some things may be very obvious, such as the

Use your senses to look for indications about where your adversary might be. Here the instructor takes a student through a room-clearing drill.

Is this a target indicator? What does this tell you about the intentions of the person on the other side? How would you handle this?

sound of a careful footfall on a creaky wooden floor, the reflection of a hidden adversary on a light fixture, a shadow on the floor in front of you, or even a gun muzzle protruding from around a corner. Target indicators may also be subtle, such as the sound of fabric scraping against a wall or the sight of an open door that was previously closed.

Do not produce any target indicators yourself. Here the author acts as a target during a dry-practice door-clearing drill.

Target indicators may be olfactory as well. For example, the smell of a smoker in a nonsmoking residence is difficult to miss. Weapon solvent, cologne, body odor, and the more primal human scents may also alert you to the presence of a hostile.

Human sign and tactical sign are indicators left behind by a sloppy adversary. Fresh, muddy footprints on a clean rug, smoldering cigarettes in an ashtray, or palm prints on a foggy glass window are good examples of human sign.

Tactical sign are any indication that the adversay has modified his environment to his presence. An open window on a cool day, furniture stacked against a door, and booby traps are all examples of tactical sign. There are others.

You may even feel the adversary's body heat as you search a close-quarters environment, such as a small hiding place. I experienced this firsthand during a search for a narcotics suspect. He'd fled our initial room assault and run to the rear of the house. After a meticulously s-l-o-w and thorough search, we reached the only room remaining: the bathroom. He'd been hiding in the cabinet under the sink for more than an hour, and I distinctly remember feeling his body heat

Do not assume that anything is clear until you have actually checked it.

emanating through the cracks in the door of the cabinet as I moved to open it. Too bad for him!

These subtle and not-so-subtle clues will be easily noticed *if* you are looking for them. They denote attempts at concealment by the target. They are hostile and dangerous indications that someone is there, hiding and perhaps waiting for you.

2. *Avoid producing target indicators.* Just as you seek target indicators during your search, you must strive not to produce them yourself. Searching a building for a hostile is 50 percent hunting and 50 percent avoiding being hunted. At such times, stealth is king! Unless you are forced to rush into a confrontation (for example, a stranger in your kid's bedroom), take it slowly, carefully, and methodically. Be quiet, be careful, move slowly, and handle each tactical problem individually. If you make an unintended noise, stop, look, and listen for about 60 seconds before proceeding.

Maximize your distance from potential danger areas—especially corners!

3. *Do not assume something is secure until you've checked it out yourself.* Do not rationalize something that is out of place; check it out and be sure. I was once searching a residential area for an ax-wielding madman who'd tried to lobotomize a couple of citizens in the best Viking tradition. I was moving along the front of a residence with my backup man when we heard a slight metallic sound coming from the driveway area. After being alerted to the sound (audible target indicator), we began moving down the driveway. Halfway to the backyard that lay beyond it, we heard a clothes dryer operating inside the house. It sounded as though someone had forgotten to empty the change from his pockets before starting the dryer. We rationalized the sound as the metallic sound we'd heard and dismissed the possibility of the villain's presence. After a superficial scan of the yard, we retraced our steps to the street. As we reached the next driveway, our boy ran out into the street, away from us, ax in hand, from the yard that we'd just "cleared"! Luckily, everything turned out fine, but don't you make the same mistake. Remember: be dead sure or be dead.

Move tactically! The purpose of any movement is to allow your muzzle to cover the danger areas you encounter, as you encounter them.

Keep your weapon in a position that will enable you to respond instantly to any threats.

Pay attention to the basics when it comes time to shoot. The best tactics in the world will not help you if you cannot hit your adversary before he hits you.

4. *Maximize your distance from potential threats and minimize your exposure to them.* Stay away from corners and any other area that you cannot see beyond as far as geography will allow. Do not let your muzzle (or feet) protrude into the unsecured space in front of you. Doing so will not only betray your position and intentions, but it may get your weapon snatched from you. It may even get you killed.

Do not allow your visual focus to wander from the direction your weapon is covering.

Except for special situations explained in the text, keep your eyes, muzzle, and potential target in line. This is the "three-eye" principle.

Part of tactical training involves using realistic targets. The author explains the "center of mass" and "threat perception" concepts during an advanced tactics course.

5. *Move tactically.* Keep your balance as you move from one problem to another. Keep your weapon in a position to fire instantly at any threat. The purpose of any tactical maneuver is to allow your muzzle to cover the potential danger areas as you encounter them. Observe the three-eye principle. This means that your weapon must be oriented toward whatever it is your eyes are looking at. Wherever your eyes go, your weapon must also go. Keep the weapon in a ready position or "hunting" attitude so that it does not obstruct your vision while you search. When moving through open areas, do so briskly but do not run unless you are already under fire. Move at a brisk walk unless approaching a specific danger area. When closing in on a potential danger area, move by using the Taylor-designed "shuffle step." Avoid crossing your feet at such times because it will impair your ability to respond in all directions.

6. *When it is time to shoot, pay attention to the basics.* My associates and I jokingly call these the "three secrets": sight alignment, sight picture, and trigger control. These "secrets" will allow you to get fast, solid hits on your adversary in the least time possible to keep him from doing the same to you. Remember, you cannot miss fast enough to make a difference. You cannot miss fast enough to win a gunfight. If you cannot hit on demand, all the tactics in the world will be of no use to you.

Whether you are a homeowner checking a noise at 03:00 hours, a police officer responding to a burglary alarm, or a SWAT member conducting a covert search, these principles will help minimize the danger and keep you one step ahead of your adversary.

THREE

BUILDING SEARCHES: CORNERS AND HALLWAY INTERSECTIONS

I will go before thee, and make the crooked places straight: I will break in pieces the gates of brass, and will cut in sunder the bars of iron.

Isaiah 45:2

One of the fundamental rules of tactics is to stay away from corners. Corners are second only to doors as a potential hazard for the searcher. If you are a homeowner, part of your homework tonight is to examine your castle for blind corners. List them all. Don't ignore items of furniture that create a "corner effect." If you can place reflective items (such as mirrors or polished lamps) in strategic locations that will allow you to look into those corners surreptitiously, then do it. This will allow you to clear these corners on the approach. If such redecorating is not likely, then you must resort to the angular search.

In years past, trainers advocated what they called a "quick peek" to clear corners. This involved sticking your head into the area beyond the corner and then just as quickly pulling it back to safety. By doing this, the operator hoped to get a brief visual "picture" of the area beyond.

There are many things wrong with this technique, and it has lost favor among the cognoscenti. Primarily, the searcher who executes this "peek" will not be in any position from which to fight and shoot. Remember, you want to be able to clear danger areas in front of your gun muzzle. The quick peek doesn't allow you to do this. Additionally, the amount beyond the corner that will be seen is quite minimal. Try it yourself. Unless Cro-Magnon man is standing right there in your face, you might miss

him completely. And what if he *is* right there? You are in no position to do anything about it. By peeking, you have programmed yourself to withdraw as part of the technique. Now you are in retreat mode, and he knows exactly where you are. What would you do if you were him? You'd attack the unprepared fool and mash his head into a Pleistocene mess with your club as he withdrew back behind the corner, that's what! Repeat after me: thou shalt not peek!

The angular search, as it is used on corners, allows you to clear the unknown space beyond the apex of the corner, incrementally, a sliver at a time.

To conduct the angular search on a corner, position yourself as far away from the corner as geography will allow. Move laterally, keeping your weapon trained on the space beyond the apex of the corner. Move toward the plane created by the far wall of the corner. Move slowly enough to be able to pick out anything that is out of place and that might be a target indicator.

When scanning for the adversary, use the vertical method of changing visual focus. Move the focus of your eyes along a central longitudinal axis, or in and out. Shift the axis slightly to vary-

The five photos on these two facing pages illustrate the angular search as used on corners. This search requires that you begin as far away from the corner as the terrain will allow. Moving laterally in a slow and controlled manner, you search the area beyond the apex of the corner.

Keep your weapon in a ready position
that is low enough to allow you to see
someone hiding in a low position.

If you face a corner at close quarters with a shoulder-fired weapon, you are better off to use the close-quarter/underarm assault position than the low ready.

ing directions as you search. This will allow visual detection of target indicators that may have initially been out of the direct line of sight. The eyes focus naturally in and out, not side to side, so it is better to search that way.

You are looking specifically for a clue that there is someone there. You might see a hat brim, the toe of a shoe, or even a gun muzzle. If there is someone there, you will likely see him long before he sees you. When this happens you can dictate the dynamics of the confrontation and either withdraw and challenge . . . or simply overrun him.

When you do decide to take the ground, do so quickly and forcefully. When conducting an angular search, you will reach a point where your angular movement will eventually allow your adversary to also see you. If it is your intent to move aggressively, you must do so before you reach the point where you have exposed yourself to the adversary's view. Conduct the angular search until you've seen what you need to see (i.e., target indicators) and then move briskly, ready to shoot, and take the ground.

The photos on these two facing pages show the author nearing the corner, conducting an angular search, and locating a hostile on the other side.

Hallways are also a potential hazard because, like any other channelized area, there is only one way through them: one way in and one way out. There may also be rooms or corners along the hallway and corners at the intersections with other hallways that may contain potential threats. They must all be cleared before going on. The end of the hallway where it intersects or bisects another hallway is also of special concern. You must keep partial attention on it as you deal with the other situations.

When moving down the hallway (or any other channelized area) keep your eye and muzzle oriented toward the potential danger area that you intend to clear next. If there is a second danger area that has not been cleared, you simply cannot ignore it.

One method that has worked well for me is to keep my eye and muzzle oriented toward the primary danger area that I've determined is to be searched next. As I close in on the danger area, I keep partial attention on the secondary danger area that

When you are operating alone and facing two different danger areas, you must eventually divide your attention between the two as you commit yourself to one or the other.

concerns me through peripheral vision, as well as taking an occasional glance toward it with both eye and gun muzzle. Before committing to finally clear the initial danger area, I glance quickly toward the secondary danger area without shifting muzzle orientation and then commit to the first danger area. At such close quarters, moving the entire gun turret (upper body, gun, and eyes) takes too long because you are at the threshold of the next danger area. You're already committed to moving in one direction; now you simply double-check the other danger area before continuing. This will help minimize the possibility of the situation's having changed behind your back. This is called "dividing attention." Is it perfect? No, not at all. But when you are alone, you have no choice.

Do not walk down the middle of the hallway; stay to one side. Avoid scraping your back against the wall, but, nevertheless, stay close to one wall or the other. Move briskly but carefully from one obstacle to the next.

What about that intersection at the end of the hallway? Quite a problem, isn't it? This is where team members and partners become very desirable. A single operator can still handle such a problem, but the risk grows substantially greater. A single oper-

Try to see as much of the area beyond the T-intersection as possible without actually getting committed to entering that space.

ator must divide his attention between the various danger areas along the hallway and the hallway intersection. Hallway intersections come in two basic types: three-way (or T-intersections) and four-way.

On a three-way intersection, you must divide your attention between the two corners and maybe what is behind you as well. When dealing with a four-way intersection, you will be presented with an additional danger area directly in front of you.

Hallway intersections are handled by conducting an angular search from one corner to the next, searching incrementally until you reach the other corner. Think of a T-intersection (or three-way intersection) as basically two separate corners opposite each other. Deal with one corner individually without breaking the plane formed by the far wall. When you are satisfied that you cannot see any farther without breaking the plane, move in a half-circle back toward the second corner, clearing it as far as possible without breaking the plane.

The main concern here is the extreme angles on either side of the hallway intersection as well as what lies farther down the

Once you are committed to entering, move diagonally along the opening briskly. Notice the quick glance to make sure your "six o'clock" is clear.

hallway. The close danger area (extreme angle) is much more of a concern than the potential danger area down the hall.

When you are committed to entry, always try to move toward your strong side if possible. You will be much faster that way. Move diagonally toward your strong side, crossing the opening from one side to the other. Break the plane with your support-side foot so that your next step will be with your strong-side foot. Move as if you intend to assault a threat hiding in the extreme angle to your strong side. As you commit yourself and actually break the plane, glance quickly toward the opposite side to make sure your back is clear.

"Holy *?!@," you say. "That's too dangerous, and it violates the three-eye principle." Yes, you are correct on both counts, but it is the best balance of risk for a single operator. Tactics must be flexible. The only alternatives to this either take much too long or ignore one of the extreme angles until entry is completed. Both of these other options are too dangerous to consider.

The immediate action drill for hostile contact is simple. If there is a hostile on the right and not on the left, the glance will only cost you one-tenth of a second because you execute it on the move. You will still be able to neutralize the adversary before he

You cannot look at two places at once, so a quick glance over your shoulder is mandatory if you are operating alone.

can react to you. If there is a threat to the left and not to the right, your glance will let you know about his presence there. Your continued diagonal movement toward the strong side will "clear" the right side of the doorway as you break the plane. Now, pivot 180 degrees toward the hostile on the left, drop into an unsupported kneeling position, and shoot him on an upward angle.

If Murphy is working overtime and there are two hostiles on opposite sides, the same technique will work. As you move to the right, your glance will alert you to the presence of the hostile on the left, and your existing diagonal rush will take you to the point of contact with the hostile on the right. Shoot the hostile on the right, at contact distance—*one time*. Immediately, drop and pivot 180 degrees into an unsupported kneeling position and shoot the second man. What are the odds that you might also get shot? Well, there is always that possibility, but remember also that this is a proactive technique. This means that you are forcing the issue and making them/him react to you. This also means that your adversaries will already be behind the power curve in terms of their reactions. They may know you are there, but not specifically when you will penetrate their space. It will take them approximately two seconds to react to your assault. If you move quickly and decisively, you actually do have the advantage.

If you do not encounter a threat to the other side, continue your forward momentum and take control of the space beyond. If there is a threat to the front, you are in a good position to neutralize it.

If you encounter a threat when you execute your rearward safety check, your response is to continue your forward momentum, but convert it into a "drop and pivot." This will allow you to operate within your adversary's reaction time and shoot him before he shoots you.

If you are starting to realize that searching alone is very, very dangerous business, then I've gotten my point across. Remember, tactics are not a guarantee, only an insurance policy to minimize the risks. But when you search alone, it doesn't get minimized very much.

Operating with a long gun does not change the concept at all. You still clear as much of the space beyond as possible without committing to enter. Be careful that your muzzle does not protrude into the unsecured area!

Commit yourself to entering and execute the quick glance.

If your six o'clock position is clear, go on.

If you locate a threat at the execution of your quick glance, convert your forward momentum into a drop and pivot. Take the long gun from your shoulder and fire from the close-quarters (underarm assault) position.

DIAGRAMS

The following diagrams illustrate the proper way to search corners and hallway intersections.

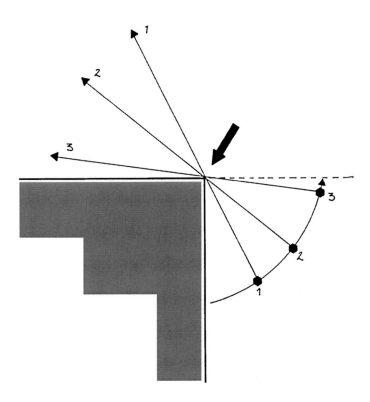

The angular search allows you to clear the unsecured area incrementally without overexposing yourself to danger.

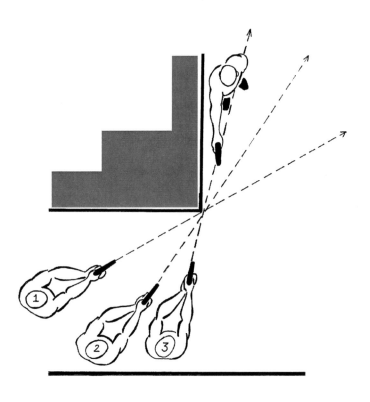

If you locate a hostile, you will be able to see him before you are seen. This allows more options than a surprise encounter would.

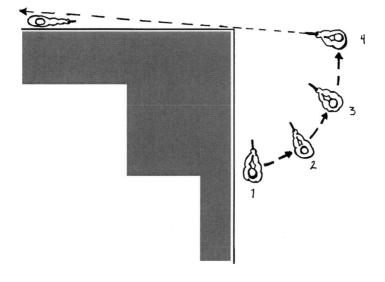

Stay as far from the corner as the terrain will allow. Realize that the farther away from the apex of the corner the adversary is, the more room he will have available in which to hide. Look deep into the extreme corner as you clear the apex.

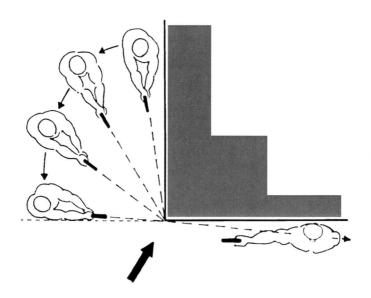

Avoid moving into the unsecured space beyond the apex of the corner. This space may be described as an imaginary extension of the far wall of the corner.

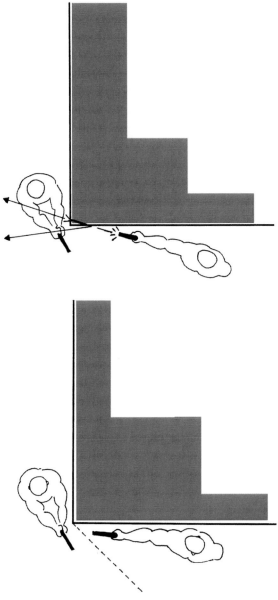

Don't crowd the corner! Not only will this give rounds greater penetration and ability to ricochet into you at the apex, but it will let your adversary know your whereabouts.

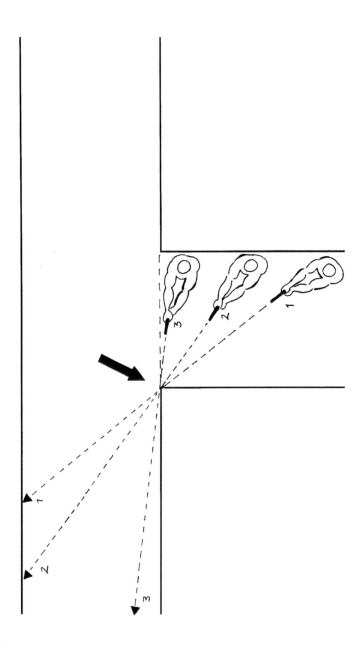

T-intersections are simply two opposing corners.

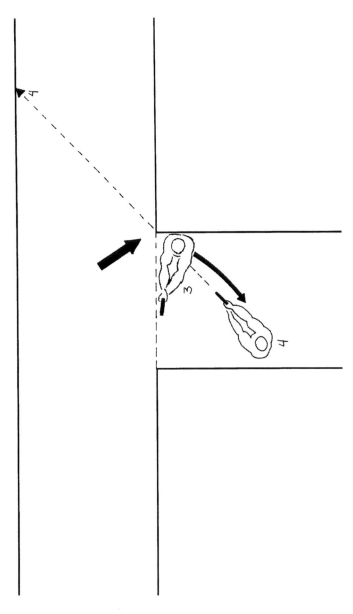

Clear one corner and then work into a position where you can clear the other corner as well.

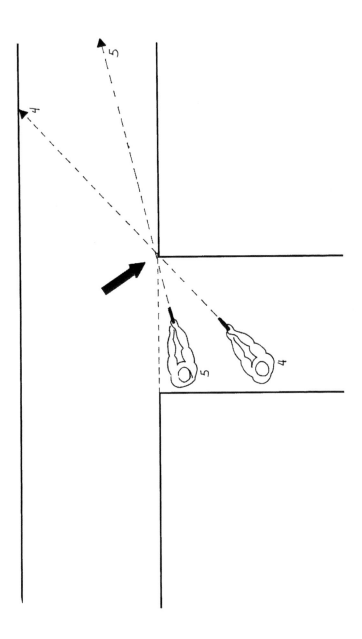

Clear as much as you can of the extreme corner without committing yourself.

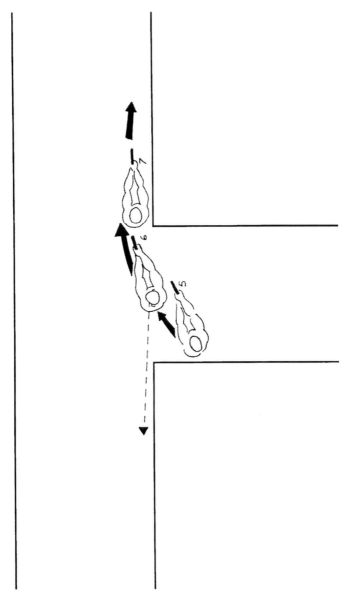

When you do commit yourself into the threshold, do it dynamically. The action begins at step 5; you clear the portal in step 6 (and take a quick glance to the other side in midstride). You then clear the extreme corner in step 7.

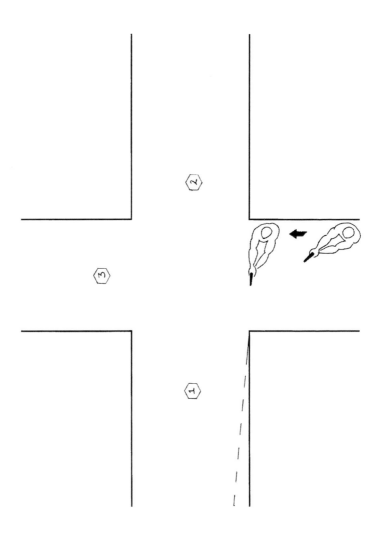

Four-way intersections are a bit more complicated because there are three danger areas instead of two.

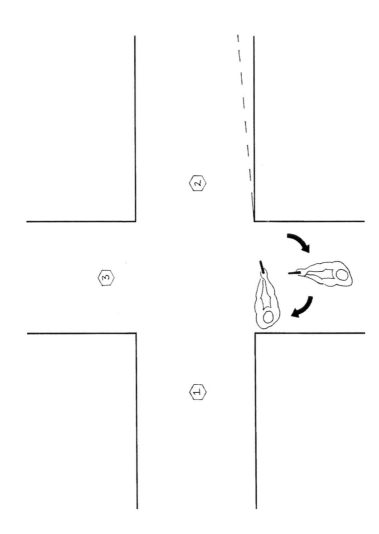

This situation is handled the same as the T-intersection, except that you must divide your attention and keep an eye on the far hallway.

FOUR

BUILDING SEARCHES: DOORS

———◆———

We shape our buildings: thereafter they shape us.

Winston Churchill

Among the most dangerous and difficult architectural features to clear are doors. Whether they are open or closed, you rarely know what is on the other side—and there is only one way in or out. Since you will likely begin your search through a door, you should pay particular attention to developing your door entry skills.

There are as many types of doors as there are types of pistols. If you are studying this text as a homeowner or businessperson, you must become familiar with the types of doors you find in your home or business. If you are a policeman or a tactical team member, you must learn about all types of doors. The types most often encountered are solid doors that open inward or outward, sliding doors, split-level doors, double doors, and spring-loaded doors.

Regardless of the type, there are certain principles to observe when facing a door. First, do not stand directly in front of the door, whether it is open or closed. The area directly in front of the door has often been called the "fatal funnel," and many who have dallied there have paid for their mistake with their lives. Note that when you clear a room visually from the outside, you will traverse this "funnel" momentarily. There is no alternative to this if you are to gain a visual scan of what lies beyond the door. It is a balance of risks, nothing more.

Doorway clearing is divided into three sections: (1) approaching and manipulating the door, (2) soft-checking the doorknob to see if it is locked or open, and (3) pulling the door briskly to the rear while stepping clear in case you make immediate contact.

Always "soft-check" the doorknob before trying to forcefully open the door. This will prevent rattling an already unlocked door and thus compromising your location.

Position yourself so that you will gain as much visibility as possible into the room beyond as soon as the door begins to open. An important consideration is to avoid leaning across the door to open it. This not only puts you into a position from which you cannot fight, but it also exposes you to substantially greater danger. The best compromise between visibility and safety is to position yourself on the side next to the doorknob. This way you will have immediate visibility into the next room if the door opens outward. You will have somewhat less visibility when the door opens inward, but that will change as you proceed with an angular search to the other side of the door.

As you start to open the door, pull your weapon back toward your chest in either a close-contact position (if using a pistol) or an underarm assault position (if using a long gun). This will do two very important things: it will prevent you from covering your support hand as you operate the doorknob, and it will allow you to shoot any aggressor who decides to burst through the door as you open it, without allowing him access to your weapon.

As the door begins to open, take a step back away from it to provide better visibility, prevent weapon retention problems, and give you room to fight. This simple step saved me a particularly nasty time one night with an overfed pit bull that was doing guard duty for a drug dealer. Toss the door open as far as it will go without necessarily slamming it. If the door opens outward, you will be able to control it and move it farther as you begin your angular search. If it opens inward, and it did not move as far as you wanted it to, you must use the door crack to gain further visibility. As you reach the other side, you may use the support hand to open the door further before committing yourself to entering.

Conduct an angular search of the room within. Remember, try to see as much as possible from the outside, before being committed to entering.

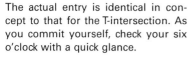

The actual entry is identical in concept to that for the T-intersection. As you commit yourself, check your six o'clock with a quick glance.

If you do not see a threat, continue in and dominate the room.

If the architectural layout prevents you from standing next to the doorknob, you must compromise and deal with the door from the other side. Such situations may occur when doors are placed into a corner or at the end of a long hallway. If the door is spring hinged, you must manage it as best you can without sacrificing your shooting position. The best way to do this is to open the door in the usual manner, and then use your shoulders, hips, or feet to hold it in position while you move.

Before entering, try to obtain as much visual intelligence about the room as you can through an angular search. The angular search (sometimes referred to as "slicing the pie") involves moving incrementally in a half-circle, clearing the room a sliver at a time and a step at a time from the outside. Use the nearer door post as a pivot point from which to conduct the search. Don't forget to scan along a vertical axis, not a horizontal axis. As you look into the room, move your visual field forward and back, altering the axis slightly each time until you are satisfied with what you've seen.

The instant of entry viewed from within. The actual check to the rear takes only a tenth of a second, and you are . . .

. . . instantly back to the original direction.

Be careful not to linger in the midpoint of the opening of the door. This is the proverbial fatal funnel, especially if the room is dark and you are in a lighted hallway. There is no arguing that you must cross this area to complete your search, but do it quickly.

In the end, the entire room may have been cleared except for the extreme angles on either side of the door. In experiments that I've conducted, I noted that an operator can see for about seven feet into this extreme angle from outside the room. Eventually the angle becomes great enough that an adversary may be able to go unseen until after entry is made.

As you quickly enter through the doorway, check these extreme angles first. Remember, you can't look in two directions at once (nobody said it was easy!). If you have a partner with you, you cut your risk in half, but that is for another chapter.

The main problem is that you simply do not know which side the adversary is hiding behind, to the right or left. This

If you encounter a threat, front sight—press! Repeat as necessary and on the move.

If you encounter a threat to the rear during your quick glance, pivot and drop. Again, front sight—press!

being the case, you must conduct a tactical coin toss. This is handled in the same manner as the hallway intersection. Remember that you are the one initiating the action. Someone hiding behind the wall in the extreme angle will be forced to react to you, thus he will be behind the reaction curve already.

There are really only three possibilities to this scenario: there is an adversary to the right angle, there is an adversary to the left angle, or there are two adversaries on opposite angles.

If there is a hostile to the left angle, you will see him as you glance to the left. Still within the adversary's reaction time envelope, you step forward with your strong-side foot and pass through the doorway, escaping the area your adversary is most likely to shoot at (the fatal funnel). Now, drop down into a kneeling position as you pivot toward the threat. This will move you out of his line of fire (in hopes of causing him to miss) and allow you to shoot upward into him.

If there is a hostile to the right, you can engage him as you clear the doorway threshold. The glance takes less than one-tenth of a second. This is well within his reaction time envelope, even if he's ready for you. He who moves first wins.

If there are two hostiles on opposite sides, there is no arguing that the odds are against you. You can still make it. Mobility and decisiveness are your assets. As you complete the glance (hostile left), continue your diagonal rush to the right. You will see the hostile on the right and engage him immediately. Continuing your "conditioned reaction," you pivot toward your support side and drop into an unsupported kneeling position. Shoot the hostile on the left. Tight spot? Damn right it is!

After you've moved through the doorway, you must slow down. Sweep the room again to be sure you didn't miss anything. You may not have been able to visually clear the entire room. There may be pieces of furniture or other things that you could not see around and which may conceal a hostile. You must physically clear them before leaving the room. Do not dismiss the room as "clear" until you've seen all four walls as well as the ceiling and floor and are certain no one is hiding there.

When you go through a door be aware of the danger areas and the extreme angles where you cannot see without committing yourself. When it comes time to enter, do so quickly and without hesitation. These tactics will help manage the danger and, with luck, they'll open many doors for you.

Once inside (if you have not made contact with any threats), turn immediately to the direction opposite your direction of entry and sweep the room for any other potential threats.

Tactics students maneuver through training obstacles with the help of instructors.

DIAGRAMS

The following diagrams illustrate the proper way to search doors.

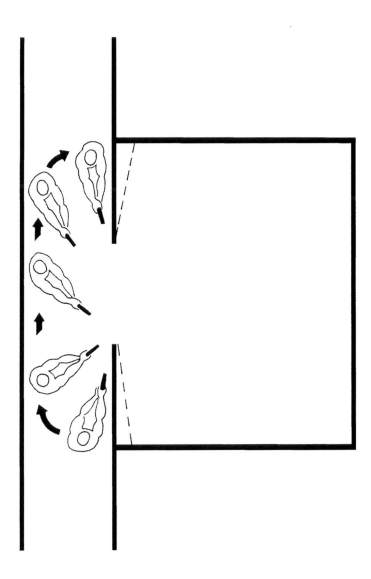

After opening the doorway, conduct an angular search of the room from outside.

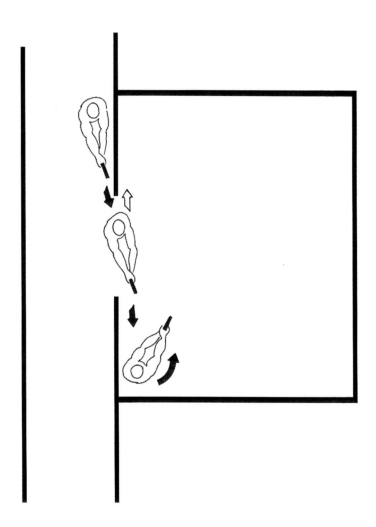

Deal with the doorway as if it were a narrow T-intersection.

FIVE

BUILDING SEARCHES: STAIRS

———◆◆◆———

I intend to go in harm's way.

John Paul Jones
Letter, November 16, 1778

Stairwells have always been a particularly dangerous problem. The main problem is that like hallways, stairwells channelize your approach. But they also present a series of simultaneous potential danger areas as well. These are the stairs' compartment itself, any corners created by switchbacks in the steps, and the upper (or lower) landing.

For many years, tactical thought on the issue was that searching down was much easier than searching up. In actuality, this depends on stair construction and design. In terms of ease of movement, it is much easier to clear upward than it is downward.

Clearing from the top (downward) forces you to expose your lower body first. It makes moving behind your gun muzzle a problem. Additionally, many stairs are not solid, and a hostile hiding under (behind) the steps can easily see you as you approach. It is much too easy for a person to do that in the area underneath the steps created by the perpendicular room offset.

There have also been proponents of the clearing style that requires operators to slither around on their backs or bellies as they negotiate the stairs. This is not only noisy (target indicators, anyone?), but it sacrifices your greatest asset—mobility.

Clearing upward allows you to clear areas behind your gun muzzle. The ideal circumstance is that your eye and gun muzzle cover each potential danger area before you expose the rest of

When you first encounter the staircase, your immediate concern will be to clear the first landing. This will be done visually from your original position.

your body to it. In the end, you must be adept at both upward and downward clearing. If you are upstairs when the Manson gang breaks into your castle, you must be knowledgeable in downward movement. Conversely, if you are downstairs when that second-story window is forced open by the Mountainside Cannibal, then you'd better have thought out the dynamics of clearing from the bottom as well.

A searcher will approach a stairwell much the same as a door or corner, searching along angular lines. He will clear as much of the initial staircase as possible before committing himself to the steps themselves.

We will discuss clearing upward first. Take note of the location of the upper landing. As you move onto the steps and begin the climb, you may be faced with the immediate corner if there is a switchback on the steps, as well as the upper landing. If the landing is directly overhead as you step on the first step, you must momentarily ignore the corner and concentrate instead on the area above you. This is where a partner is very useful indeed.

As you commit yourself to the steps and break the plane created by the staircase, you must cover the upper landing as well. The best way to do this is to place your back toward the solid wall (but not actually touching it) and cover the upper landing from below with your weapon. Clear this problem by using the angular search—a slice at a time.

Now, your partner can deal with the staircase and any switchback corner while you hold the landing. When he's done, he'll hold the upper landing while you move up and clear the next danger area. This is teamwork.

If you must solve this problem alone, the way to minimize the hazard is to move sideways up the stairs, keeping your back to the wall (but not actually touching it). Keep your weapon aligned with the potential danger areas. Since there are two or perhaps more of these danger areas, you will be moving your weapon back and forth in an attempt to cover them all in succession as you move.

As you begin to ascend the steps, you must turn slightly and move your search upward to visually clear the upper landing.

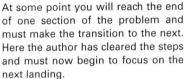

At some point you will reach the end of one section of the problem and must make the transition to the next. Here the author has cleared the steps and must now begin to focus on the next landing.

When you visually clear the upper landing, you may continue to take ground. Notice that the author is actually ascending in a backward mode, always facing the danger area.

As you reach each switchback, you must treat it as if it were a corner. Use the concept of the angular search.

Clear the next landing as you approach it. The key to a multiple-danger area, such as a staircase, is to deal with one problem at a time.

Do not cross your feet as you climb or descend. Crossing your feet will compromise your mobility. If your feet are crossed in midstride and a hostile pops up on your dominant side, you will not be able to respond to him in time. Use a sideways shuffle instead. With the sideways shuffle, you are able to pivot to the strong side or support side as well as move behind cover or even abandon the staircase quickly.

Stay back from the "well effect" at the center of a multi-flight staircase. A hostile hiding at the top or bottom can easily fire on your position without being detected. This central well portion allows a person at the bottom floor to see up to the top and vice versa. If you can see that far, so can the man you are hunting—and he can shoot you too. Just as with the other geographic obstacles we've already discussed, handle each portion of the problem individually, move carefully, and be ready to shoot at all times.

Descending is done using the same concept. In fact, depending on the design of the stairs, descending is sometimes easier. However, you must be careful that you do not expose your feet to what's below.

Conduct an angular search around the first corner and clear the space from above.

DIAGRAMS

The following diagrams illustrate the proper way to search stairs.

A stairway presents various danger areas simultaneously. After the initial flight of steps has been cleared, you must begin clearing upward and may be required to ascend backward.

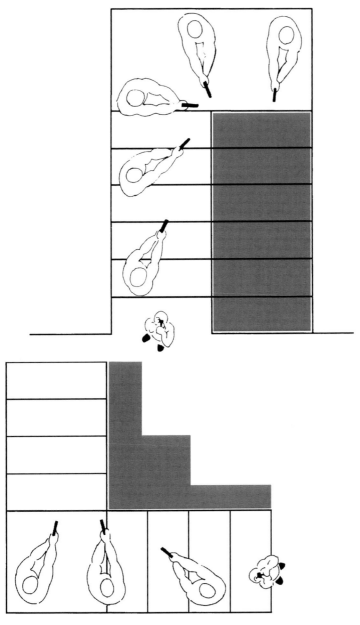

Staircases vary in their construction and design. The key points to remember are to divide the visible danger areas and to handle them one at a time.

SIX

BUILDING SEARCHES: EXTREME CLOSE QUARTERS

———◆◆◆———

Small rooms or dwellings discipline the mind; large ones weaken it.
Leonardo da Vinci

All structures will present some extremely close-quarters situations that must be dealt with. These places are particularly dangerous because while negotiating them you are literally trapped in them. All of your immediate mobility is gone.

Windows are one of these often ignored tactical problems. Avoid window entries if you can. I've used window entries when stealth was important, but I also had several operators pointing their guns into the room from other windows to keep any hostiles away. Windows are convenient gunports for your adversary when you are outside, so do not walk in front of them. Windows are also fatal funnels when you are inside.

Because windows are not generally considered to be entry points or hiding places, they're often ignored during a search. This might be a life-altering mistake. An officer I know was conducting an interior search for a robbery suspect who'd surreptitiously exited the building and hidden himself just outside a window. When the officer crossed in front of the window, the suspect fired at him from outside. Surprise!

The same thing can easily occur in reverse, and you could take fire from a building through a window. While you can easily see out through a screened window, you cannot see in. This effect is accentuated when there are varying degrees of light and darkness inside and outside. The bottom line: be careful with windows.

No matter what type of weapon you are equipped with, if you expect to operate in extreme close quarters, a handgun is hard to beat.

Pistols are small, maneuverable, and extremely portable—attributes that make them very attractive for close-quarters engagements. The only thing that they do not have is power.

If you encounter an obstacle that does not allow you to maintain control of your pistol, do not try to force an outcome this way. Do you think that this shooter is in a position to shoot if a threat materializes?

A much better option is to move to the close-contact position. This keeps the pistol close to the body to prevent muzzle protrusion or an unexpected weapon retention problem.

If the danger area is too close for a standard ready position, but not close enough for a close-contact position, you may elect to simply "compress" your ready position enough to maintain control of your muzzle.

There are also extremely close-quarters spaces that may require checking. Such places include attics, basements, storage closets, and, in some instances, extremely cluttered rooms that are often found in depressed urban areas.

Attics are not good places to go for the lone operator. Tactical teams use dogs and portable shields to search an attic. As you break the horizontal plane of the ceiling through the attic opening, you are surrounded on all sides with potential hiding places for your adversary. Not good. Consider this: Why would you ever need to search by yourself in the first place? If you suspect that an adversary has hidden himself there, he is in no position to attack you, and he is obviously on the defensive. Hold your position and call for reinforcements. The same goes for crawl spaces below the building. If the unthinkable happens and the adversary begins firing up at you through the floor or down through the ceiling, you can do the same to him. But whatever you do, do not follow him into these places.

Some environments prevent the use of any weapon but the pistol. This shooter is about to clear a crawl space in an old aerospace research facility. Close-quarters environment? You bet!

Clearing small rooms and closets is also problematic, because you do not have room to maneuver. You may need to manage the door with your body weight and alter your shooting platform to see what you need to see.

Bring the weapon into the close-contact position when you must clear such areas. This shooter is clearing a rest room stall. Rest rooms are a favorite hiding place for criminals who are trying to go undetected.

When you search a storage closet or any extremely close-quarters environment, your main concern should be weapon retention. Do not move your weapon into an unsecured space or plane. If there is a hostile hiding there, he will snatch it away from you and . . . well, you know the rest.

At such times, you must either retract your ready position so that your pistol is closer to your chest in a "tight Weaver position," or you must abandon the two-handed hold and use the close-contact position. This involves holding the pistol in a position so that the wrist is braced tightly alongside the pectoral muscle. The pistol is aligned on the target/danger area and canted slightly outboard to avoid snagging clothing or equipment during firing. Your support hand and arm are held forward about chest high, but away from the muzzle, to assist you in moving or to deflect a hostile's first strike.

These positions are quite efficient within arm's reach (up to about four feet). Any shots fired in this environment will probably be within inches of your adversary, and he will likely have powder burns on his clothing along with a bullet hole or two. You will be able to get reasonable hits to the body of your adversary up to approximately five meters using this position.

SEVEN

TACTICAL MOVEMENT TO CONTACT AND SHOOTING ON THE MOVE

The "Art of War" is simple enough. Find out where your enemy is. Get at him as soon as you can. Strike him as hard as you can and as often as you can, and keep moving on.

Ulysses S. Grant

Being able to shoot as you move from one point or obstacle to the next is very important. The entire reason and purpose for any tactical maneuver, whether slow or quick, is to allow you to cover the next potential danger area with your gun muzzle. Sometimes, you must also shoot on the move as you advance toward another position. You may even wish to aggressively close on an adversary to shoot him on the advance, put him on the defensive, or enable you to execute a more precise shot.

As you search, you must adhere to the three-eye principle. This means that, except for specific situations, both your eyes and your gun muzzle must always be oriented in the same direction. They must, all three, be "looking" at the same thing. If you look up, your muzzle "looks" up also. If you look at the apex of a corner for a target indicator, your muzzle must be oriented toward that corner as well.

The objective is to be able to shoot any suddenly appearing hostile without hesitation, at all times—whether you are stationary, advancing, or passing him on the move.

When you are moving through an open area, you should use the tactical walk, where you avoid exaggerating any of your movements or body mechanics. Walk just as you would walk down the street except for a few accommodations to the firearm you hold. The weapon will be up and oriented toward the point

The most important thing to remember is that the shooting platform exists from the waist up. What the lower body and feet are doing is almost inconsequential to shooting.

Advancing aggressively toward an adversary may be required in a tactical scenario.

you are approaching. Your knees will be slightly bent, and you will have a slightly aggressive forward lean. This in turn will help stabilize your upper body and allow a sort of rolling gait. The movement looks as if you are gliding, without any up and down motion, from one point to the next. Take smaller steps than you normally would to keep control of your feet and motion as you move. Most important, as one U.S. Marine Corps close-quarters battle instructor pointed out to me (as only a marine can), move only as fast as you can guarantee getting hits on the hostile. Do not break into a run unless you are already under fire because you cannot shoot effectively "on the run."

Moving forward is not the only technique you must learn. It may be necessary to move directly from one point to another while covering your flank with the muzzle of your weapon as you move.

Covering a potential danger area or shooting to your support side as you move along a straight line is easy; all that is

When you decide to engage, do so dynamically. Control the action—do not let it control you.

required is for you to pivot your body at the waist to the support side and point your weapon as you need it.

Covering a similar danger area or shooting to the strong side on the move is slightly more difficult because of the inflexibility of the human torso and the upper-body dynamics of most shooting positions. I've conducted some experiments to determine which method best solves this problem. The method I think is best, and which I use, is to turn the body enough to allow a comfortable shooting position toward the danger area. This will place you more square to the target than a "usual" Weaver stance. Now you simply move in a sideways shuffle as you cover the danger area with your weapon.

The most important thing to remember is that the shooting platform exists from the waist upward. What the lower body and the feet are doing is almost inconsequential to your shooting. Isolate the shooting platform from its transportation (i.e., the legs).

These same concepts are as useful for evacuating an area as they are for "assaulting." If you are in a situation where you want to get out of the area quickly but do not want to turn your back on danger areas, you can simply walk backward while keeping

When closing on a potential danger area, do so with stealth and the ability to respond instantly. The Taylor-designed shuffle step will allow this.

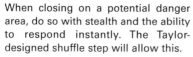

your weapon oriented toward the danger. Instead of walking "heel-toe," you reverse it and walk "toe-heel." This allows you to feel for unseen obstructions behind you as you move.

As you approach a particular danger area, such as a door or the apex of a corner, you must slow down considerably and alter your gait to the Chuck Taylor-designed shuffle step.

The advantage to this method is that it allows you to approach very carefully and slowly without compromising your stealth or ability to shoot. This method also discourages you from crossing your feet as you walk, which adversely affects your mobility in tight quarters as well as your ability to respond to various angles.

The shuffle step is equally useful for moving forward and backward as it is for moving laterally. It is similar to the type of footwork that would be employed by a Western boxer. The leg that is closest to the desired direction moves first and is followed by a catch-up step with the other leg. Beginning with a slightly bladed Weaver stance, the footwork patterns are as follows:

To move forward: Take a half-step forward with the support-side leg (front leg) and then catch up with a half-step with the strong-side leg.

To move rearward: Reverse the process and begin with a half-step rearward with the strong-side leg (rear leg), and then catch-up with a half-step with the support-side leg.

To move laterally right or left: The process is identical in concept.

To move to the strong side: Step to the strong side with the strong-side leg a half-step, and then catch up with the support-side leg

To move to the support side: Reverse the process for the strong side.

If you must close distance quickly, the tactical walk will allow you to do so while retaining the ability to shoot en route.

Engaging a hostile to the firing side is somewhat more complicated and involves a slight change in footwork.

Engaging a hostile to the support side is easily accomplished on the move without altering the normal stride.

When performing the shuffle step, do not allow your feet to touch at midstep. Keep a slight distance (one-half shoulder width) between the feet on each step.

A second common mistake is dragging the feet. Do not fall victim to this. Step with the toe first and then allow the heel to make contact with the deck. Remember that stealth is of the greatest importance when you are using this technique.

Along with stability and enhanced potential for quick movement, this type of footwork allows you to clear an area using the angular search method. The tactical walk and the shuffle step both have a valuable place in an ace tactician's "war bag"—so learn them.

EIGHT

TACTICAL USE OF COVER AND CONCEALMENT

I have known men in the west whose courage could not be questioned and whose expertness with the pistol was simply marvelous who fell victims before men who added deliberation to the other two qualities.

Bat Masterson
"Famous Gunfighters of the Western Frontier"
Human Life Vol. IV, 1907

One of the primary lessons a gunfighter must learn is the difference between cover and concealment. If you learn and use these tactics well, they will be invaluable during a fight—and may very well save your life. If you use them poorly or confuse one with the other, you may become a candidate for that most dubious of honors, the "Distinguished Wooden Cross."

In a nutshell, cover is anything solid that offers ballistic protection. That means that you can hide behind it and be reasonably certain that bullets fired at it will not penetrate and hit you. This requirement may be problematic depending on the type of armament your opponents have on hand. Concealment, on the other hand, is anything that hides your presence from the adversary. A good example of cover is a brick wall; a good example of concealment is darkness. Brick walls will stop most small-arms ordnance. Darkness will allow you to hide, but it will not stop anything. Sometimes cover will also offer concealment.

Cover may be used during a fight to prevent your getting shot. Concealment may only be used prior to the fight to deceive your adversary about your whereabouts. Concealment will provide a base from which to launch a surprise attack. Concealment demands stealth, which is sometimes enough. If your opponent doesn't know you are there, he won't think of shooting you. Cover often offers the same advantages as concealment with additional ballistic protection.

To use concealment you must have prior knowledge of an enemy's approach, as well as the belief that he hasn't seen you. Obviously, if he knows you are there, hiding in the shadows will not help you. To use cover you must have the time and room to reach it, as well as the specific anticipation of hostilities.

There have been discussions about the propriety of always running to cover when a fight begins. Generally the closer the fight is, the less time you will have available to respond to a threat. You will often be forced to react and solve the problem with your own gunfire. Remember the nature of most urban gunfights: short-duration, high-intensity fights. If you are being attacked, your response must be to counterattack immediately. Then, after you've reacquired control of your environment, you can move to cover. Sometimes, however, the gunfire may come from an unknown area or from such a long distance that it precludes an instant counterattack from you. At such times, you must move to cover first. A good rule of thumb is that if you do not have a target to shoot at when you come under fire, *get behind cover.*

Realize that many things most people consider to be cover are really only concealment. When I conduct tactical courses, I often place a hostile target directly behind a couch with only his gun and eyes showing. Many students take valuable extra time to place that precise brain shot on the exposed target area when they could simply fire through the couch (concealment not cover) and drill him in a third of the time . . . twice! Similarly, most of the things people tend to hide behind, thinking they are taking cover, can easily be penetrated by gunfire.

The same goes for corners, doors, and walls inside buildings. Most modern cartridges will penetrate directly through these light wood and stucco structures. Therefore, if an adversary fires at you from a doorway, you can shoot him right through the wall. Even buckshot will work in such situations.

Automobiles, on the other hand, make relatively good cover against most small-arms fire except for centerfire rifle fire. The exception here is the side glass, which is as resistant to gunfire as a piece of paper.

You want to seek as hard a point of cover as possible, but realize that such hard cover will tend to cause projectiles to ricochet. Bullets often ricochet along an axis parallel to the cover they've struck and angled slightly away from it. If you are too close to your cover, they'll ricochet right into you. The magic distance seems to be at least six feet. If you stay at least six feet from your cover, the angle of departure of the ricochet will have grown enough to bypass you completely. If you crowd your cover, you run the risk of getting hit with one of these ricochets.

Remember when I said that there are no absolutes to tactics and that tactics are an art and not a science? Remember when I promised in the introduction that you would see this again? There are times when you may wish to get closer to cover than the six-foot standoff distance. This is when your adversary is shooting at you from above, such as from a second-story window. If you move too far from your cover, he will be able to bypass your cover by virtue of his higher position and shoot down into your position. Yes, by moving closer you may run a greater risk of ricochets, but you run a greater risk of getting shot outright the other way. Additionally, if you are engaging multiple adversaries at different points, you must be cognizant of the possibility that they may pinpoint you and attempt to flank your position. So note your adversary's location and stay alert.

When you are looking around cover in a search or shooting around it, expose as little of yourself as possible. Use a roll-out technique so that the only things exposed to the potential danger or the known hostile are your gun muzzle and your eye. Avoid changing hands to shoot from the support side. Remember, the reason you are rolling out with your pistol is to shoot and stop the so-and-so who is shooting at you. I know very few men who even approach their dominant-side accuracy when shooting from the support side. The purpose of shooting is to hit. To roll out to the support side, turn your weapon 90 degrees to the support side so that it is held in a sideways position. Now you can roll out with much less exposure than if you were holding the weapon in a standard position.

Shooting around cover is preferable to shooting over cover. Here a student rolls out around cover only enough to be able to fire at the target. Notice his distance from the cover itself.

Do not use the cover to "brace" your shooting platform unless the target/threat is a great distance away. The target in this photo is barely visible 85 meters away.

Taking a position this close to the cover may allow ricochets to skip right into you.

A better technique is to stay at least six feet away from the cover you select. Sometimes this may not be possible, but it is always preferable.

The architecture did not allow this operator to increase his distance from his cover, but notice how he exposes only enough to be able to see and shoot. Notice also the rolled-over position of his pistol. This is much more preferable to the old method of switching hands.

The instant you clear your cover, you must be able to shoot.

Learn different shooting positions
that will allow you to fully use your
cover. Here a tactics student perfects
his kneeling position.

Cover and concealment are very useful tools to have and
often will become crucial factors in the future of your breathing
activities. Study them hard and keep your bases . . . well, covered!

The only available cover is low, so this shooter must lower his firing platform to conform to the available cover. (Photo courtesy of Chuck Taylor.)

What is this shooter using, cover or concealment? If it does not stop bullets, it is not cover!

NINE

REDUCED-LIGHT OPERATIONS

———◆◆◆———

Night brings our troubles to the light, rather than banishes them.
Lucius Annaeus Seneca
(c. 4 B.C.–A.D. 65)

A great percentage of conflicts in urban settings occur after the sun goes down. Although most urban areas in the United States are always lighted to one degree or another, learning to manage low-light tactics should be on everyone's must-do list.

The problem in reduced-light environments is identifying targets. If you can see well enough to recognize an attack, you do not need any sight enhancements to solve the problem. In such cases, you simply shoot as you do in the daytime. When the daylight or ambient light begins to dissipate enough that your weapon's sights begin to blend into the background, as well as on the target, you must rely on the inherent reflexive muscle memory of your Weaver stance. This is the level of reduced light where the popular radioactive tritium sights are at their best.

Almost everyone these days has some form of tritium sight affixed to his weapon, whether it is a pistol or a submachine gun. These sights are useful to a certain degree, particularly during that brief time between last light and full-blown night. This is when there is still enough light to see an adversary, but there is also enough darkness that distinguishing your sights from the target mass is virtually impossible. Of these sights, I have found that the three-dot variety serves best for rapid alignment on targets. Some operatives do not have the option of tritium sights because of budgetary or political reasons. Don't laugh. Once

Operating in extreme low-light environments requires the use of artificial light. All weapons that are expected to see low-light action (such as this Smith & Wesson pistol) can be equipped with dedicated-light mounts.

Low-light operations primarily require target identification. If there is enough ambient light to see a target you may do without artificial light.

Colt 1911 pistol with Sure-Fire dedicated-light mount. (Photo courtesy of Laser Products.)

when I was training a group of California police officers in the techniques of low-light shooting, two officers from a reasonably progressive department told me, in complete seriousness, that they were not authorized to install "radioactive night sights" on their weapons because their city had been declared a nuclear-free zone! One disadvantage with tritium sights is that they stand out very brightly in the dark and may give away your position.

Shotgun dedicated-light mount on a Benelli Super 90.

New-generation pistol dedicated-light mount on a Smith & Wesson Sigma pistol. (Photo Courtesy of Laser Products.)

New mount for Heckler & Koch's USP series pistol. Laser Product's mount is substantially more robust than the factory-issued unit. (Photo courtesy of Laser Products.)

Even rifles can benefit from the addition of a white light unit if they are intended for close-quarters use. (Photo courtesy of Laser Products.)

Beyond this light level, a flashlight of some sort is essential to locate and identify your target. There are many types of lights on the market these days that are as bright as anyone will ever need for an interior or exterior search. Keep them simple. You do not need a light with colored lenses, multiple buttons and switches, or bells and whistles. A simple bright, pressure-switch-operated, focused-beam light is all that's required. A good rule of thumb is to get a light as small and as bright as you can find. (For a source guide on lights, check the appendix of this book.)

There are many methods for incorporating a flashlight and a firearm. The ones I've found to work best are the highly specialized integral weapon mounts most often found on SWAT weapons. However, although an entry team can cart these lights around on their weapons, the street cop or private citizen will generally not want to do so. Their option is to use one of the bright compact flashlights in conjunction with a modified shooting grip on the weapon.

When using a pistol, the flashlight technique (conceived by Mike Harries) works best. For those using long guns, the compact flashlight can be adjusted for sensitivity and held alongside the weapon's fore-end. Held in this way, it is activated by grip pressure. (For a more detailed discussion on this subject, study the appropriate chapters in my previous books, *Tactical Pistol* and *Tactical Shotgun*.)

The main thing to establish when using a flashlight in conjunction with a firearm is rough coaxiality between the axis of the weapon's bore and the light beam. You don't need exactness in the axis between the light beam and the bore, since the target will even be illuminated with residual light. This in turn will allow you to see your sights superimposed on the adversary. If your light beam is adjustable, set it on a focused beam (not a wide-angle-spread, diffused beam). Once you've set the proper hold on the light and have the light and weapon roughly aligned, you may begin moving and searching.

Initially your light will be off. You will only turn it on to scan an area that you simply cannot see clearly enough with the ambient light. Do not stroll through the combat zone with your light

The Harries flashlight ready and firing position is the best method for integrating pistol and light. It is both stress- and recoil-resistant.

beam on as you search. Remember that a beam of light tells any-one who is interested where you are. All your adversary needs to do is to shoot toward the light. Move tactically, just as you would normally move during standard light conditions. Use stealth as much as possible. When you reach an obstacle that you cannot clearly see, bring your weapon up, light the area, look for hostiles, shoot them if necessary, turn the light off, and move carefully to another spot. Do not track the light along the deck from your position to the danger area and back. Wait until your weapon is oriented on the danger area before turning the light on.

If you are illuminating the apex of a corner as you clear it, you must light from the bottom corner of the apex wall upward. If you keep your weapon up all the time, you might miss an important target indicator near the floor, such as the toe of a shoe or even an adversary lying prone.

If you are scanning a large room or a room from a doorway before entering the room, you have the option of scanning with the light along a horizontal axis—in essence, sweeping the room with light. Once one sweep is executed, move carefully to another position and repeat the process. The main difference between the corner and the doorway (and even the T-intersection) is that with a single corner you have one potential danger area, whereas with a room or hallway intersection you have many. That is why you must sweep the area instead of focusing the beam on a specific danger area. The same goes for stairways.

If you encounter a hostile, keep the light on him—preferably on his eyes so that his vision and reactions will be impaired. Obviously, if he is armed and you are out in the open without cover, you'd shoot him. If you do not need to shoot him, then you must place him in a situation that gives you reasonable control over his movements. Order him to keep his hands up, get flat on his stomach on the floor, etc. At such times, keep the light directly on him. If you allow your light to go off, and he decides to attack you at that moment, you will not see him do it until it is too late.

If the hostile turns out to be an innocent party, shining your light in his eyes will conceal your gun pointed at him. If necessary, you can even keep the light pointed toward his face while you reholster. He will never know you had him "covered."

Realize that the instant your light goes on two things will happen. One is that your "night vision" will diminish. If you "light-scan-turn off-and-move," your eyesight will not have adjusted to the change in light enough to be able to see very well in the dark environment. One technique that I've tried with varying degrees of success is to close my shooting eye whenever I turn the light on. You are scanning for hostiles, not shooting. You can scan just as easily with your other eye as you can with your master shooting eye. This way, when the light goes off, your shooting eye will still have some degree of night-vision capabilities. If you encounter a threat while scanning, it is a simple matter to open the other eye as the pistol intercedes the line of sight and you reflexively look for the front sight. In any case, at that point, shooting your adversary will supersede the preservation of night vision.

The second thing that will happen when you illuminate an area is that anyone hidden within that search area will be instantly advised of your presence, thus eliminating the element of surprise! That man hiding in the room or around that corner may choose that very time to attack. You must be ready to shoot whenever your light is on. Remember the issues of cover and concealment. If you can scan an area from behind cover, then do so. If your adversary thrusts his weapon blindly around the corner or around whatever he is hiding behind to shoot you, beat him to the punch and shoot right through his cover.

Searching alone in the dark is more dangerous than searching alone in a lighted environment. Darkness really demands reinforcements. If you are fortunate enough to have your partner (or partners) with you, the problem is eased considerably. Your partner may light up an area from cover and keep his light on. A hostile hidden in the danger area will either attack immediately or hold his position in hopes of going undetected. If he attacks, you and your partner will shoot him. If he holds his position, he will probably do so as long as the light is on him. While the light is on him, he will be unable to see anything except the light. In this situation, you may approach under and outside the light beam and clear that position without being seen.

What about gun-handling problems? Do they change at all when the lights go out? Reactive gun-handling manipulations, such as emergency reloads and malfunction clearances, depend on the operator's ability to diagnose the specific problem with his weapon. There has been some discussion in tactical training circles that you should simply execute a tap-rack-flip as a conditioned reflex anytime your pistol fails to function. One well-published trainer suggests simply reloading the pistol anytime you experience a malfunction. Such simplistic approaches are certainly attractive to disciples of the KISS (keep it simple, stupid) school. The only problem, however, is that weapon stoppages (empty-pistol slide locked back included) must be cleared by specific means. You cannot "fix" an empty pistol with a tap-rack-flip. Neither can you clear a feedway stoppage with a reloading procedure. To do so often exacerbates the problem, creating a worse stoppage than you had initially.

This is why the symptom-solution approach to malfunction clearances works best, regardless of lighting conditions. The problem of low light, however, may complicate the determina-

To clear a malfunction in low light, secure the light under your firing-side arm (lens to the rear), fix the problem, and then retrieve the light.

tion of what condition your weapon is in if it does not fire. If there is enough ambient light for you to determine the need to fire (without artificial means such as a flashlight), then you will also be able to visually determine the status of your weapon. If it is so dark that you cannot see your weapon in front of your face, then your adversary will not be able to see you either. You can use the darkness to your advantage . . . and concealment. The need to fire in a low-light environment will probably be determined because you have illuminated a "threat." Your illumination will also reveal the condition of your weapon. If you experience a malfunction under such circumstances, the best course of action is to extinguish the light, move to an alternative position immediately, and clear the malfunction in the darkness. If you cannot see the adversary, he cannot see you. You can use the darkness as concealment while you determine the status of your pistol by feel and get it back into action.

If you experience a failure to fire, you will know instantly by the sound and feel of the weapon, no matter what the lighting conditions. A tap-rack-flip maneuver will fix the problem posthaste. A failure to eject is solved with the same maneuver, but you must be certain that you actually have that particular stoppage and not the similar-looking feedway stoppage, or empty gun. You can determine what you have by visual examination using available light. In no-light conditions, you can use your support hand to feel for cartridges, empty cases in the ejection port area, or partially ejected magazines, for example, to determine what you have. I cannot stress enough that, in no-light conditions, your first reaction to a stoppage must be to extinguish the light and move. Then you can fix the problem.

If you must shoot an adversary, do not turn the light off immediately afterward. Remember that you must verify the result of your gunfire. If you experience a failure to stop or (perish the thought) you missed, you need to know that. The only way you will gain that information is to keep the light on him after shooting, as you lower your weapon to the ready position. I know that someone out there is screaming, "That will illuminate you as well, and other bad guys will know where you are!"

A Laser Products dedicated-light fore-end for a Remington 870 shotgun.

Special operations weapons, such as this suppressed MP5, can be enhanced with the addition of a white-light unit. (Photo courtesy of Laser Products.)

An often forgotten aspect of low-light operations is muzzle flash. Excessive muzzle flash will destroy your night vision as well as signal to everyone where you are. Test your ammo for muzzle flash. If it is excessive, replace it.

That potential threat may be true, but the real threat is the one you've just fired on. Remember, tactics only minimize, not eliminate, danger. This situation calls for a balance of risk. Making sure that the real threat is neutralized takes precedence over possibly giving away your position.

Solving any tactical problem in the low-light or no-light environment is very difficult. Study the preceding text and the following diagram and then practice the concepts. They will prove to be an illuminating experience when you go hunting at night.

DIAGRAM

The following diagram illustrates the proper way to search in reduced-light operations.

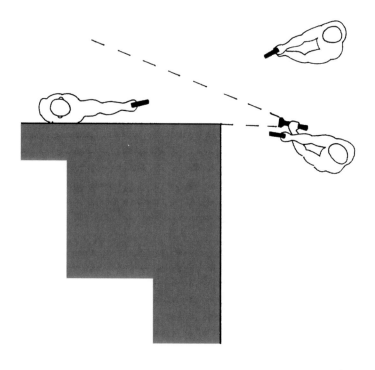

Operating as a team, your partner may light up an area while you move outside the beam of light and clear the area surreptitiously. For all purposes, you will be invisible while the light is on.

TEN

TEAM TACTICS

———◆◆◆———

Ask for volunteers for dangerous work. Pick out the best. Train them in fellowship. Then they will develop qualities that no one has ever suspected them to possess. They will follow you through anything—they will even live and fight and go on to certain death by themselves.

Waffen SS Col. Otto Skorzeny
"The Father of Special Operations"

Clearing and searching a building is considerably safer and easier if you have extra people with you. This chapter is based on the operational characteristics of a three-man element, or team. A three-man team includes a contact man, a cover man, and a security man. For purposes of clarity in the diagrams at the end of this chapter, they will be designated shooter A, shooter B, and shooter C, respectively. Using this concept, shooter A, the contact man, is the primary searcher. He is the first one in line, and the one who actually "clears" each obstacle. Shooter B, the cover man, provides assistance to the contact man. He helps the contact man search and is an extra pair of eyes for situations where there are two potential danger areas in opposite directions, such as the extreme angles found on opposite sides of doorways and T-intersections. The third man, shooter C, is a security man. He covers the team's back.

The three roles may be interchanged during a search, depending on the dynamics of the architecture. With such a three-man team, it is possible to clear an entire building in relative safety. The team can actually move as a gun turret and cover all potential danger areas simultaneously.

The security man is also the most expendable position on the team. This means that if there are only two men present, they may still be able to conduct the operation, but the

Tactical operations, such as clearing and searching a building, are much simplified by a specially trained team. This group of three was a searching element in a tactical team. Left to right: Investigator Al Acosta, the author, and Officer Mike Hurt.

cover man must serve double duty and provide rear security as well.

The members of such a team need not be SWAT experts, but they must have a basic understanding of tactical principles as well as having established a method of communication among themselves. This concept is very workable for both law enforcement scenarios with patrol officers and for civilian security operations.

Let's examine the architectural obstacles studied earlier and clear them with a three-man team. Read the explanations and peruse the diagrams at the end of the chapter to get the overall picture of the techniques involved.

DOORS

The team members will move to one side of the door as they approach. If the geography allows, the team members will position themselves on the side nearer to the door knob. The con-

Special teams need special equipment to do the job safely and efficiently.

tact man and the cover man will cover the door area itself. The security man will cover the area of the hallway beyond the door. This configuration is called "stacking."

The contact man will initiate the movement across the door as he begins to conduct his angular search into the room from the outside. As he moves, the cover man moves with him, maintaining physical contact with him at all times. The cover man protects the contact man's flanks and back down the hall as he executes his angular search. It is important for the team members to know which areas each man is responsible for. If the cover man and the security man are covering the same area, one of them must secure another area. Eventually the cover man will be committed exclusively to the hallway beyond the door. At this time, the security man (on the other side of the door) will focus on assisting the contact man as he moves through the door. At this point the cover man and the security man will actually exchange roles.

The actual movement through the door is initiated by the contact man. The security man (now acting as cover man) will key his action on the action of the contact man. The contact man enters the room low along a diagonal line through the door and immediately checks the extreme angle on his side of the door.

His security man (now acting as his cover) enters the room as simultaneously with the contact man as he can. He moves to the opposite side, also along a diagonal line to the extreme angle on his own side of the door. He moves in a higher posture than the cover man so he can enter the room almost over the top of the contact man. The result is that both extreme angles are covered almost simultaneously. Each man will look deeply into the extreme angle first and then sweep along the wall inward toward the center of the room.

As the first two men enter the room, the man covering the hallway will key on their movements and also move into the room. He doesn't give up his area of responsibility (i.e., the hallway). He maintains his focus on the hallway and simply moves into the room just enough to use the doorjamb for cover. When the two men inside the room have completed their search, they will "stack up" behind the man covering the hallway. They will exit the room and proceed down the hallway. In this case, the man who had been covering the hallway during the room search will now become the contact man and the man behind him his cover. The last man out automatically becomes the security element.

HALLWAYS

If the area behind them is not secure, the security man will be moving along with the other two men, but facing backward as he covers the area behind them. If, on the other hand, the area behind them is secure, he may also bring his weapon and his focus forward along with the rest of the team.

The team moves down the hallway in a cloverleaf formation. This requires the three men to move in a triangular configuration. The contact man is in front, and the cover man and the security man are to his immediate right and left. They are in physical contact with one another to facilitate nonverbal communication. All three guns are oriented down the hall. The contact man moves his muzzle along with his eyes as he searches. The other two men have crossing and overlapping zones of fire. This means the man on the right side of the hall is focused

Tactical ballistic vests, weapon-mounted lights (such as these MP5-mounted Sure-Fire units), and trained teamwork are essential factors in tactical operations.

Properly trained team members support each other's tactical movements. Here two team members conduct a dynamic entry in a training environment. (Photo courtesy of Chuck Taylor.)

toward the left side and vice versa. All three men may shoot without endangering each other.

Their guns are in the "hunt" position. The muzzles are not held so high that the sights obstruct their view, but neither are they pointing at the deck. It is important to be trained to lower the muzzle quickly after shooting in order to observe the post-injury actions of the adversary. This still occurs . . . *after* you shoot. When you are searching and hunting, hold the muzzle slightly higher, usually covering the area you are searching. Sometimes a team member must cross your line of fire to reach a position from which he can cover a danger area. In such cases you must dip your muzzle as he passes in front of you to avoid covering him.

CORNERS

The contact man approaches the corner with his cover man. The security man covers both of them. The contact man again moves slightly lower than the cover man to allow him the room to shoot over him. The cover man is directly behind the contact man and moving with him. As they clear the apex of the corner, they assume the cloverleaf formation again and proceed with the search.

T-INTERSECTIONS

Searching this very hazardous architectural feature is again facilitated by the use of the three-man team. As the team approaches the intersection, the contact man and the cover man will turn away from each other in a back-to-back formation. They rely on the physical contact from each other for nonverbal communication. Remember, the contact man calls the shots. He is the one who will initiate any movement toward or away from an objective. The security man may cover the area behind them if it is not secure. If it is a four-way intersection and the area to their rear is secure, he will focus on the hallway in front of them.

When the contact and cover man have conducted their angular search, they will both commit themselves to the extreme

The Heckler & Koch MP5, here in its personal defense weapon format, is the first choice for indoor tactical team problems.

angles simultaneously from the center of the hallway. They will have determined which way to proceed (right or left) before-hand, and the contact man will, of course, have searched in that direction. When the extreme angles are clear, the security man will move up behind the contact man and become his cover, while the cover man holds down his side of the intersection. When the team resumes moving, the cover man now becomes the security element again as the team moves down the hall.

STAIRWAYS, CLEARING UP

The main thing to remember with a team stairway search is to avoid overcrowding the stairway. As the team approaches the staircase, members clear the first flight of steps with an angular search. The contact man will cover and hold the first landing at the top of the steps. The cover man now takes on the contact man's duties, and the security man becomes his cover. The cover man (now acting as contact) clears the overhead landing from below. At this point, one man is covering the first flight of steps and the upper landing. A second man is covering the overhang directly overhead, above the staircase. The security man moves forward and clears the corner of the first switchback and the upper landing from the switchback. The first man joins him, and then the man covering the landing from below begins to move up the stairway toward them, still covering the landing. When he reaches the other two men, the original contact man and his cover move up to the top of the landing and clear the area there. The last man holding the upper landing from the first switchback holds his position until he is advised by the others that it is clear.

STAIRWAYS, CLEARING DOWN

If the team is executing a downward stairway clearing, the contact man clears and holds the first flight down from the upper landing. The cover man captures the second flight from the overhang above and holds his position to allow the other two men to descend. Taking the same care as when searching upward and being alert to exposing their feet and legs, the contact man and the security man move down the steps.

Depending on the configuration of the steps and the likelihood of exposing the lower body on descent, the security man may elect to get on his chest and look into the room below through the steps. This is only an option when he has others covering his back, because his mobility is totally gone.

The two men will take the steps and the bottom of the stairs, and then they'll be joined by the third man. When they've reached the bottom, they again have three divergent zones of fire and can continue with their search.

ACTION ON CONTACT

When any man on the team confronts a hostile with either verbal commands or gunfire, he automatically becomes contact man, and the team's actions will be keyed to his tactical requirements. The others must hold their areas of responsibility. If the contact man needs assistance he will request it from one of the others. The

The nature of a tactical problem may require creative measures. World-famous tactical trainer and writer Chuck Taylor and the author conduct a second-story entry during a special-operations course.

Two tactics students move through a simulator drill "dry" to develop team movement skills.

cover man is always the first to help him (usually the man closest to him). At this point the security man must not only cover his area, but also the one abandoned by the cover man.

CASUALTY EXTRACTION DRILL

The possibility of a team member being hit always exists. It is a wise team that plans an immediate action drill for such eventualities. The casualty extraction drill is usually structured as if the first man got hit, but it is wise to vary the wounded man's location in the line of three during practice.

If the first man is hit and goes down, his cover man will step over him and place direct controlled gunfire on the source of the hostile fire. If the hostile has hidden himself behind a wall, a door, or a piece of furniture, the cover man shoots right through it. The security man moves up, grabs the fallen man, and drags him back behind the first item of cover or the first corner that is available. When the security man begins his extraction, he yells, "Moving!" When the cover man hears this, he begins moving back along with the two other men, still shooting into the hostile's position to cover their withdrawal. This procedure will work to extract the wounded man from the initial kill zone. After this, the team can evacuate the casualty and call for reinforcements. The important point is that if one of your people is hit, you reflexively shoot back at the source of the gunfire. This does three possible things: (1) it makes the adversary think about his own flanks and the incoming gunfire, (2) it may disrupt his homicidal plans long enough to get your people out of his trap, and (3) it may also cause his demise. The bottom line is that *thou shall not abandon your wounded in the kill zone!*

TEAM EXTRACTION DRILL

There are times when you may wish to abandon a building because the adversaries in it are a bigger problem than you expected or the interior requires more personnel. Proper and safe extraction requires a trailing security man to remain behind to deal with any following hostiles. As the first two men retreat behind the first obstacle they advise the security man, who has stayed behind to cover their move. The best signal is to say, "Clear." The security man acknowledges by saying, "Moving." He moves back and passes beyond the point of cover being used by the first team to his own point of cover. He takes a position from which to cover the team's next rearward movement, and announces, "Clear." The process is continued until they all reach their objective. Think of the extraction drill as a reverse leapfrog process.

RELOADING AND MALFUNCTION CLEARING

Team members may be used to support each other during weapon manipulation drills. Malfunctions and the requirement to reload will always occur under fire, so the ability to keep shooting and (it is hoped) hitting is paramount. So is the use of cover. If there is no cover immediately available and a team member experiences a malfunction, he calls out, "Red." This tells his partner that he needs "cover" to clear his weapon, and he needs to move out of the kill zone without letting the hostiles know his status. The partner will move in front of him and take his area of responsibility. Together they move to cover. Their partner covers the maneuver and, if necessary, shoots any threats. When the first man has cleared his malfunction, he calls out, "Green," telling his partner that he is back in the fight.

Multiplying your forces also multiplies your survivability. If you can bring more people to the party than the bad guys, do so. It will pay off when the bullets fly. To quote Confederate Gen. Nathan Bedford Forrest: "Get there first with the most men."

DIAGRAMS

The following diagrams demonstrate the proper way to clear obstacles using team tactics.

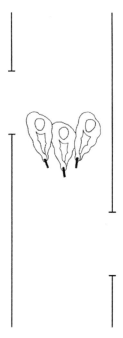

Moving down a hallway, the three men of a tactical element maintain constant physical contact with one another and have mutually supportive zones of fire. This is called the cloverleaf formation.

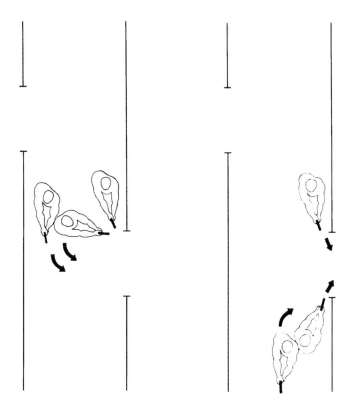

As they approach a doorway, the second and third men peel off from the formation. One man conducts an angular search of the room, while the other man covers the far hallway. The first man covers his half of the room.

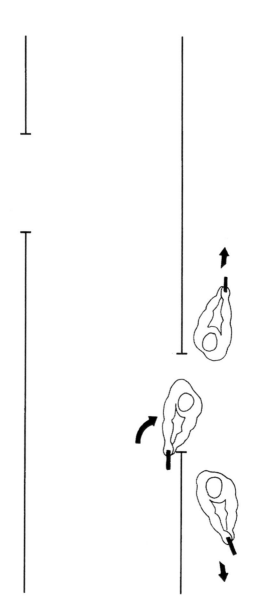

Entry is made to opposing corners in a crossover manner. The man covering the hallway backs up into the room and uses as much cover as he can while maintaining visual surveillance of the hallway.

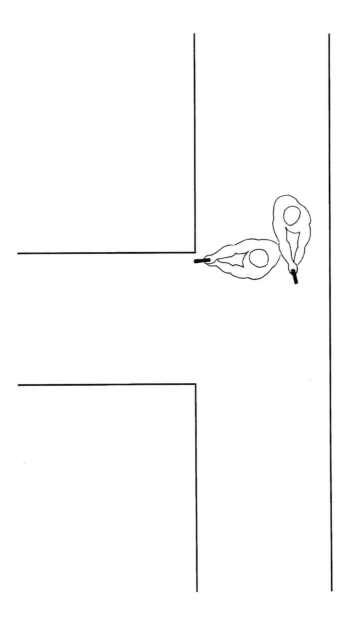

The L-shape formation is useful when danger areas are located at 90-degree angles to each other.

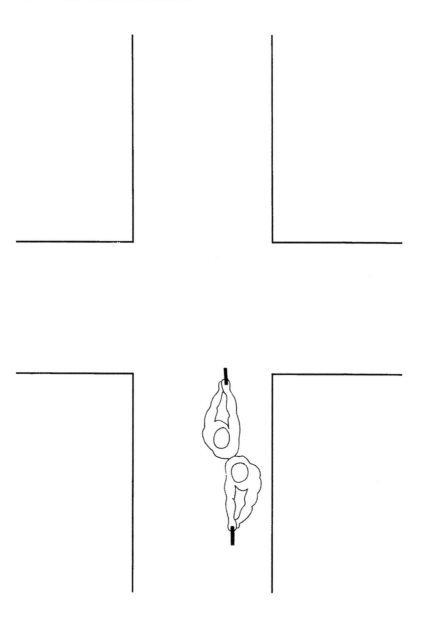

The back-to-back formation is useful for danger areas located 180 degrees from each other.

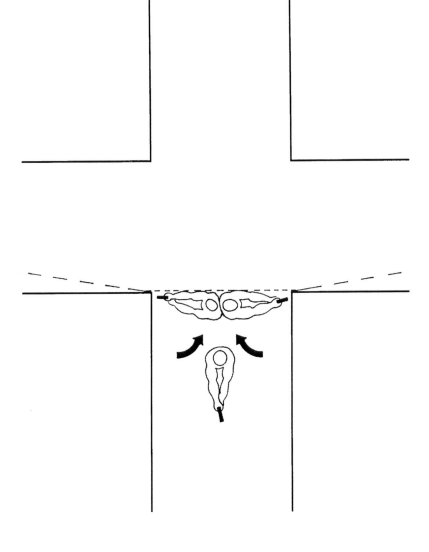

Team negotiation of the four-way intersection. Two men clear both corners while the third man covers their rear. The primary searchers periodically glance at the hallway beyond to make sure it appears clear.

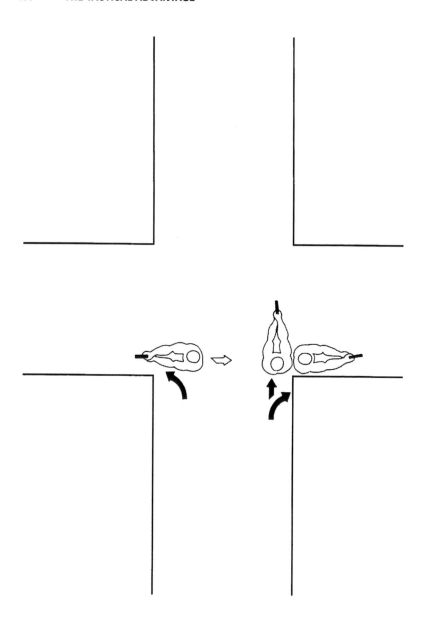

The team clears the extreme corners and covers the unsecured hallway beyond.

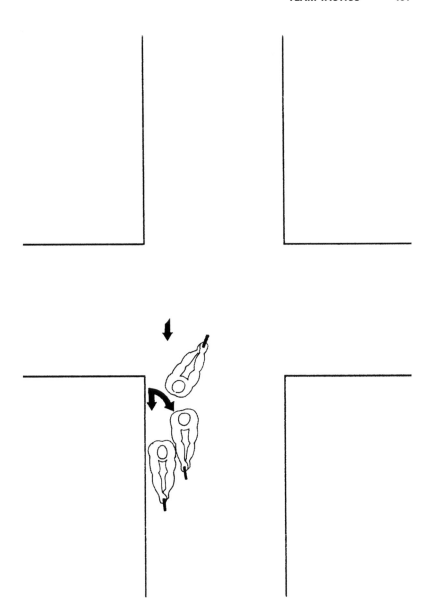

Team members resume their cloverleaf formation, slightly modified by having one team member covering the rear, and resume the search.

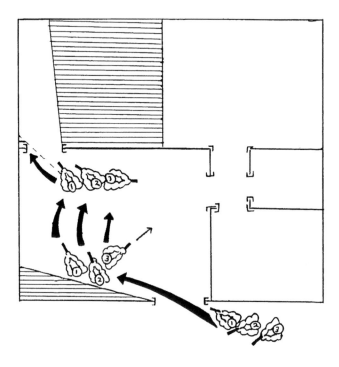

These diagrams depict a three-man assault on a single-story apartment. This is a diagram of an actual operation. Follow the flow of the team as its members maneuver through the apartment. Notice that the team maintains its unified integrity and that at no time is an operator left alone in an unsecured room.

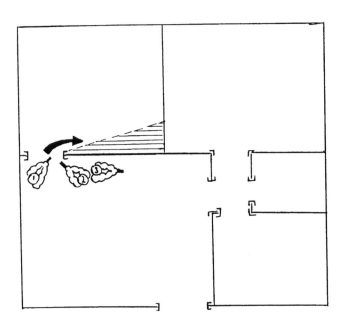

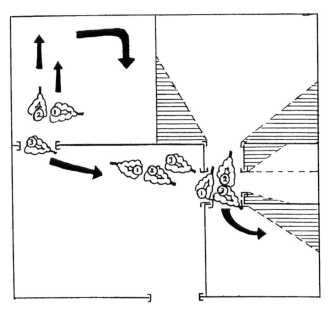

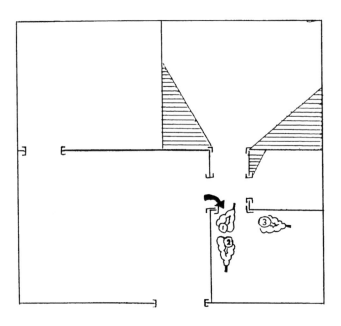

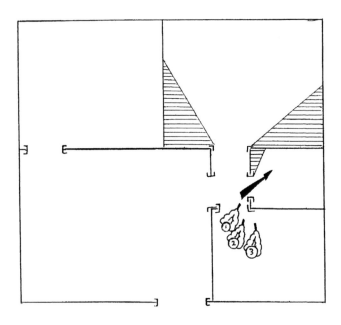

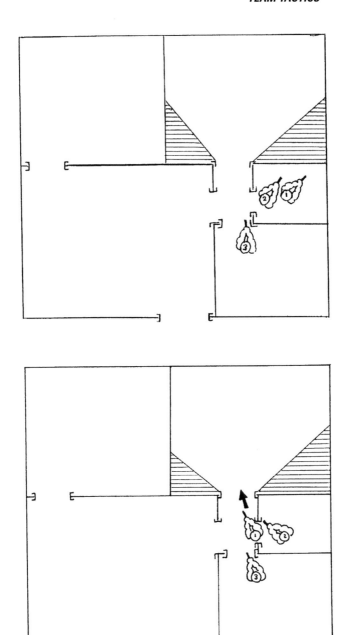

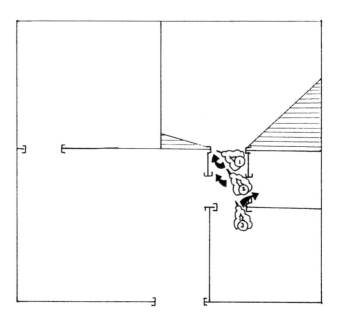

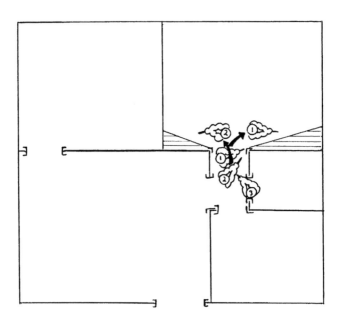

ELEVEN

TACTICAL COMMUNICATIONS

Much unhappiness has come into the world because of bewilderment and things left unsaid.

Fyodor Dostoyevski

Stealth is a primary concern when moving through a structure looking for an adversary. This dangerous situation is made less so when you have some associates helping you out. But having help also requires communication with that help. You obviously don't want to debate about the process you will use to negotiate that next T-intersection, nor do you want to announce to your partners in your command voice that there is an adversary around that corner. The information must, however, still be relayed. This is done through nonverbal communication.

The primary form of nonverbal communication is to be in physical contact with your partner so that you can "feel" which way he is moving and what area he is covering. If either of you sees something of concern, it is a simple matter to grab his shoulder or belt to stop his movement. When moving in this manner each of you knows exactly where the other is, and you can mutually support each other. When abstract things must be relayed, hand signals must be used. The hand signals I've include here are universal. The actual hand signal itself is not important, but all team members must know what a particular signal means. A full-time tactical team may require an entire alphabet of signals for nonverbal communication, but a two- or three-man team only needs a few basic ones.

"Attention!"

ATTENTION

To call your team's alert to something, raise your support hand to eye level, palm out, fingers together. After you have your team's attention, you may point at what you want them to see.

"Clear!"

CLEAR

After checking and clearing an area, show a thumbs-up indicating to the others that it checks OK.

"Cover!"

COVER

Bring your support hand palm down on top of your head or simply bring it palm down in front of you. You may also point at the area you want covered after the initial signal is acknowledged.

DANGER AREA

Using the support hand, execute a slashing motion across your throat and then point to the possible danger area.

"Danger area!"

"Hold!"

HOLD

Using the support hand, make a fist and hold it up near your shoulder-neck area. (This may also mean stop.)

"Listen!"

LISTEN

You may have heard an audible target indicator that other team members have not. To point it out to them, you can signal a "hold," followed by cupping your support hand around your ear, as if you were trying to listen.

"Hostile!"

HOSTILE

If you see a target indicator such as the toe of a shoe or a hat brim, you can alert your team by pointing in the direction of the hostile, using your support hand, with all five fingers spread wide and pointing in the direction of the hostile's hiding place.

It is paramount to use the support hand for all of the nonverbal communications. Keep your weapon oriented to the danger area in the event that you are interrupted in midsentence. These simple signals can be learned by all team members or partners. Street patrol personnel will find them extremely useful as well.

TWELVE

HANDLING NONSHOOTING CONFRONTATIONS

Never trust much to the honor of prisoners. Give them no liberties which might endanger your own safety or afford them an opportunity to escape. Nine out of ten of them have no honor.

Gen. David J. Cook
Hands Up: Or Twenty Years of Detective Life
in the Mountains and on the Plains

Now, just what do you do with a hostile when you locate him during a search? The answer most of us provide is basic and simple: if he is a threat, shoot him. But I must tell you that for every search that finishes with gunfire, there are hundreds that do not. You must be aware of managing those situations when you do not shoot. Those are the times when you've located your adversary, but the situation does not allow gunfire. You are holding him at gunpoint, and he may be exhibiting varying levels of cooperation or antagonism. What now?

A confrontation where your mental trigger does not get automatically "tripped" is probably more dangerous than the search that brought you to contact in the first place. Why? Because many operators relax their guard when their mind perceives the confrontation as a nonshooting situation. They tend to go into administrative capture mode instead of staying in combative hunter mode. The primary thing to keep in mind is *do not relax too soon!*

When you find yourself in this situation, you must do three things in succession:

1. Take control of the problem.
2. Disarm the adversary.
3. Secure the adversary.

Taking control of the problem begins by making the hostile react to you, not the reverse. If you see him first, take the initiative. If there is no immediate threat, you should still get behind cover in the event your perceptions are wrong. The position you select must allow you a good view of the adversary as well as allow you to shoot him, if necessary.

Since you are initiating the contact, you can also dictate the dynamics of it. Be certain that you are in a good position before you even consider announcing yourself. If you make contact unexpectedly, before you can get behind cover, keep your weapon pointed at him—ready to shoot. Then move behind cover as soon as tactically possible (more on this later).

If he has a firearm in his hand, you have a shooting situation. Some politically correct law enforcement trainers will undoubtedly take exception to this, and advocate *always* challenging such an adversary to "drop it." To prove the silliness of this notion, try the following experiment:

Face off with a training partner at about seven meters apart. Each of you will simulate a pistol in your hand by extending the index finger (the cops-and-robbers special of our childhood). You, as the captor, will point your "gun" at your partner. He will have his "gun" by his side like a bad guy would when you've surprised him. You will order him to drop his gun or put his hands up. On hearing your order, he will bring his hand up and simulate shooting you as fast as he can—BANG! You try to beat him to the punch by shooting him when he begins his move. Ready, GO! Pretty close wasn't it? In fact, about 80 percent of the time, he wins. It's a simple matter of action time versus reaction time. If he initiates the action (shooting at you), you are already behind the reactive power curve. No matter how fast you are, your reactions will not be quick enough to prevent his actions. Now ask yourself this: "Is the adversary's life so important to you that you are willing to risk your life to keep from killing him?" Your answer will determine your philosophy about such confrontations.

Once you are set in a dominant position (without your adversary's knowing it, of course), you may issue a verbal challenge. If he turns quickly, be ready for the gun in his hand. I believe it is foolish to allow a man to turn in your direction under such circumstances. Remember, action always beats reaction: if he turns with the intent of shooting you, he will invariably get a shot off before you are able to react. That is the best argument for always challenging behind *cover*. However, not allowing a hostile to turn toward you is still a good idea. With this in mind, the first words out of your mouth should be a loud and menacing, "Don't move!"

Your weapon should be oriented toward your adversary, but lowered enough to be able to see what he is doing. This will allow you the quickest response if your "capture" turns into a gunfight. Do not point your weapon directly on target, because you cannot see. Do not point your weapon at the deck, because you cannot shoot. After the "Don't move!" he will either submit or attack. If he attacks, you must kill him. If he initially submits, stay ready for his possible attempt to turn the tables.

Once you have him covered, stay in command of the situation. His only option is to obey your orders or get shot. Do not open negotiations with him. If he persists in jabbering away, order him to be silent. You can't blast him for it if he keeps talking, but be ready for the move he's trying to conceal with his chatter.

After the initial "Don't move!" command, pause for a couple of seconds. Stay ready to shoot if his reaction to your command is anything other than submission. The next command (that's right, command—not request) is intended to disarm him if any weapon is within reach. If he has a weapon in his hands and he's facing you . . . well, you know what to do. Keep your words short and clear. "Hands up!" will bring his empty hands up toward the ceiling and away from the waistband area and pockets, where most people stash their armaments. "Turn around!" will place him in a position from which it will be substantially more difficult for him to react to you simply because he won't see you. "On your knees!" "Hands behind your head!" "Interlace your fingers!"—these last orders will place him in a situation from which even a kung fu master would have difficulty attacking.

Now your partner can move up and handcuff him, or you can call for reinforcements via radio or telephone. If you do not have a phone on hand, you may have to move the adversary to a location from which you can call. If it is necessary, have him move on his knees. If he tells you it hurts him to do so, tell him you will end his pain forthwith if he doesn't MOVE IT!

If you are a homeowner and you are calling the police, be careful about what you say, because all 911 lines are taped. Tell the dispatcher that you are holding an intruder at gunpoint. Give his description and then give a description of yourself. Have the dispatcher repeat this to you, so the police officers will be clear about who is who. When the blue suits arrive, do not greet them with your gun in hand. Have your weapon out of sight. Conceal it in your belt or stand it by the wall next to you. The idea is to be able to respond to the threat you've captured, but also not appear to threaten those you've called for help.

Expect to be treated as a "suspect" until the reality of the situation is determined. The officers don't know you, and you would be surprised how many people habitually lie to the police. All they really know is that an alleged good guy has the drop on an alleged bad guy. Be understanding, polite, and cooperative. You would act the same way in their shoes.

There is a third possibility in an adversary's actions when he is challenged: he may simply turn and run. The court case of *Tennessee vs. Garner* prevents us from shooting a fleeing criminal simply to prevent his escape. There may, however, be situations that seem to override this. If his act of initial escape is simply an attempted ruse to buy space and time to reattack or if the attempted escape coincides with a simultaneous attack (U.S. Marine Corps Lt. Gen. Chesty Puller's "attacking in reverse"), you can shoot him. For example, if he runs 10 feet and then stops quickly and turns around to shoot you, you can shoot him first. Also, if he turns around to shoot as he's running away, you can shoot him first. There may be other factors affecting your decision about whether to shoot him, such as his activity prior to flight. What if this heathen had just brutally killed a restaurant full of people, or if he just slashed everyone in your family? Are you going to let him go? This is a touchy legal-moral situation that requires careful thought and consideration. So think and consider!

Take control of the situation, disarm your adversary, and secure him. Do those three things from behind cover. Do them forcefully with clear and menacing commands. And be ready to shoot him every second. That is the formula for handling nonshooting situations.

THIRTEEN

WEAPON RETENTION AND COMBAT COUNTERMEASURES

———◆—◆—◆———

Never hit a prisoner over the head with your pistol, because you may afterward want to use your weapon and find it disabled.
Gen. David J. Cook
Hands Up: Or Twenty Years of Detective Life in the Mountains and on the Plains

One of the most important rules of tactics is to maximize your distance from all potential danger areas or human threats as much as the geography will allow. This is a good thing to keep in mind, but it is also sometimes difficult to implement. The very nature of modern society demands that you must sometimes move in close proximity to a danger area. Clearing the interior of urban structures, for example, particularly demands close-quarters work.

One of the biggest concerns for someone operating in such an environment is the possibility of a hidden adversary's mounting of an unexpected physical attack at close quarters. Such an attack often leads to the person under attack being disarmed or even killed. There is even the possibility of encountering a non-shootable adversary who is nonetheless uncooperative and potentially combative.

Each one of these situations has its own solution. Let's discuss the issue of weapon retention first. When you move in a close-quarters environment, your weapon should be in some form of ready position. Only idiots search for adversaries with holstered or slung weapons. Other than wariness, the best solution to weapon retention is prevention.

Normally, the best position for close-quarters operations is the muzzle-depressed, or ready, position. When you maneuver

The speed rock is a reactive close-quarters technique that is used when the operator cannot move to gain standoff distance because of the terrain. (The speed rock is discussed and depicted in more detail in the next chapter.)

close to a potential danger area, however, you should retract your ready position so that your weapon is closer to your body. When armed with a pistol, this position is the close-contact position. This involves keeping the pistol close to the firing-side chest area, yet aligned on the potential threat. You can also add the support hand to the grip and simply pull the firing position in closer to the chest. You may fire on a close-quarters threat from the close-contact position or, if there is room, simply extend your arms toward the adversary to complete the Weaver stance.

If you need one hand free to open doors or to move something, your weapon goes to the close-contact position. The support hand simply relinquishes the grip and attends to its other duties. This is also the case when you are moving through very close quarters, particularly in the dark. Your support hand is out in front and slightly to the side to feel your way or deflect physical attacks at contact distance. An attacker must reach all the way to your chest to get at your gun. By that time, you will have caused him to change his plans.

When clearing similar areas with long guns (such as shotguns, carbines, or submachine guns), simply switch temporarily to the underarm assault, or close-combat position. In tight quarters, it

Practical application with training guns.

If you find yourself struggling over your firearm with a hostile, be as violent and as vicious as you can to win the fight. Remember, if you lose, you die!

Many close-quarters problems involve weapon retention. There is only one reason why someone will attempt to take your weapon!

So let him have it, instead of wrestling with him . . .

. . . and then let him have something else!

is sometimes preferable to the indoor ready position. The close-combat/underarm assault position allows you to keep your muzzle oriented directly on the danger area as you clear it. It will also give you more control with long-barreled weapons if you must release one hand for other duties. As soon as the particular close-quarters area has been negotiated, resume the shoulder-mounted low ready position. The final word is that firing a weapon from the shoulder is more accurate and therefore preferred.

These modifications to your ready position will make it exceedingly difficult for someone to pop out of a hiding place and snatch your weapon. Prevention is better than any cure, but making it difficult for an adversary to get close to you does not mean that it is impossible.

Statistical analysis of "gun grabs" tells us that the vast majority of disarmed individuals end up getting shot by their disarmers. It is prudent to assume that the man who tries to grab your gun has the unfriendliest of intentions toward you. You are also quite justified in dealing most harshly with him.

Your first reaction when a hostile grabs your weapon must be to get the muzzle pointed toward him. Your intention is to shoot the attacker, not just to get your gun back. Don't wrestle, don't try a wrist lock; rather, shoot the @#&*! Note that if he has a solid enough grip, you may not be able to force the weapon around to point it at him. You must then move your body around using the muzzle as a pivot point. If the grab occurs while you are in a compressed ready position, your reaction will be quicker and stronger. If the attack occurs when you are in a low ready, you must drop your body down low and simultaneously pivot the muzzle up toward him. If you are attacked from the side, you may need to sidestep as well as drop down.

In some cases, a self-loading (semiautomatic or selective fire) weapon may fail to eject the spent cartridge case after the first shot has been fired because the adversary's hands may have covered the ejection port or impeded the bolt (or slide) from operating. Therefore, it is imperative that you hit with the first shot—you may not get a second one. If you are armed with a long gun and you cannot manage the leverage required, you might con-

Close-quarters activity may also mean simply moving someone, either a friendly or a passive resister, out of the way.

sider letting him have your long gun and then using your pistol to solve the problem with two or three well-placed rounds.

There are times when an adversary might not present an obvious enough threat to warrant a lethal response from you. He might be unarmed, but still be aggressive. Your response to him depends on the disparity of force existing at the time. If you are a six-foot-tall, 200-pound, power-lifting, SWAT-trained, ex-Special Forces karate expert, then shooting the 100-pound teenage burglar who says he's going to kick your ass would probably not be seen as reasonable. However, if you are a 75-year-old grandmother with a bad back, no one will question your use of lethal force. You must use good judgment to determine what is "reasonable force."

Regardless of the disparity of force present, you must keep control of the situation. If you cannot shoot, and the attacker or intruder is not obeying your commands, hold your ground. Do not get in close proximity to him. If he begins to close on you, and the disparity-of-force situation discourages your shooting, you must use physical violence to control him. This situation is quite prevalent in urban police work where the bad guys know exactly what the officer's rules of engagement are and believe that they can close the distance without getting shot.

Some weapons are robust enough to be usable as contact weapons. If your weapon is fragile, avoid this.

Chuck Taylor was a pioneer in the development of realistic close-quarters defensive techniques with pistols and shoulder-fired weapons.

If you have a partner with you, one of you can close and physically subdue the adversary. Do not try this alone! If the problem escalates, your partner can shoot him as you move clear. If you are alone and you get into a physical contest with him, you've lost your advantage—specifically the fact that you could have shot him.

One good alternative here is the use of an oleoresin capsicum ("pepper") spray. While pepper spray is not the solution to all nonshooting problems, it sure beats a few rounds of fisticuffs with an adversary of unknown skill. If the encounter does come to blows, hit him fast and hard. This is also a time when you must know your limitations. This is no time for fancy artistic moves—stay with the basics. Hit him hard in such vital areas as the eyes or throat. If you kick, do so *only* below the waist. With all due respect to my martial arts friends, kicks above the belt are relatively useless in a real fight. A fast, hard kick to the leg or knee is extremely devastating and much more effective than a spectacular jumping spin kick. Meanwhile, keep in mind that you have a gun in your hand. If he gets near it, shoot him. Keep it in the close-contact position and keep your trigger finger clear of the trigger. Remember, you are not sparring with him. You *do not* want to be that close to him for very long. Your intention is to hurt him so badly and so quickly that he either submits to your will or his thoughts turn from attack to self-preservation and he retreats. In basic terms, the more you hurt him, the less he'll think about hurting you. This may sound barbaric, but who cares? It works!

If you are armed with a shoulder gun, your empty-hand options will be limited. In this instance, you can use the weapon itself as a striking implement. For example, you can thrust with the muzzle to vital targets as you would with a bayonet-equipped rifle. Rising or vertical strikes with the buttstock are also useful techniques. You can even strike with the receiver in a horizontal manner to the chest or throat of an adversary.

Engaging in physical contact while you are under arms is to be avoided at all costs, because the risk of losing your weapon is too great. The dynamics of personal tactics involved in building searches, not to mention daily life encounters, demand getting close to potential adversaries. When that happens, these combat countermeasures will help you save the day.

Close-quarters confrontations may require unusual firing positions. For example, how would you respond to a threat if you were seated in a vehicle?

FOURTEEN

DISTANCE INTERVALS AND CLOSE-QUARTERS DEFENSE

———◆–◆–◆———

It is the cold glitter in the attacker's eye not the point of the questing bayonet that breaks the line. It is the fierce determination of the drive to close with the enemy not the mechanical perfection of the tank that conquers the trench. It is the cataclysmic ecstasy of conflict in the flier not the perfection of his machine gun that drops the enemy in flaming ruin.

Gen. George S. Patton, Jr.
Warriors' Words

The closer you are to your adversary, the fewer options you have at your disposal. There are roughly three distance intervals involved in close-quarters defenses:

1. Within grasp of the adversary, or face to face, and with no room to maneuver or gain standoff distance
2. Within arm's reach of the adversary, but with the ability to gain standoff distance
3. Just outside of arm's reach, one or more steps away

When you are within the grasp of the adversary or he has already grabbed you, there is a great immediacy to solve the problem . . . right now! Every instant that you are within this distance your life is in jeopardy. You will not have room to extend your pistol into a standard firing position, nor will you have any room to maneuver. Often, this is because your back is against a wall or vehicle, or you are in a phone booth, or whatever. This will impede your ability to gain any standoff distance prior to firing. Your only option in this situation is the Taylor-designed "speed rock."

The speed rock involves going to step three of the presentation (clearing the muzzle from the holster and rolling your shoulder down, thereby allowing the muzzle of the pistol to

This is the closest interval—where you are literally within your adversary's grasp.

Common or traditional pistol techniques will not work if your back is trapped against an object or if you are physically grabbed.

move forward along an arc toward the target) and then "rocking" back at the knees. This rocking motion, done in conjunction with the upper body movement of the presentation, will orient the muzzle onto your opponent's chest. Two quick shots may be fired in less than one second. In fact, I time students with an electronic timer, and the average time, including the reaction to the audible "start" tone, is .89 second. The fastest I've seen is .75.

After the speed rock is executed and the shots are fired, you do not want to linger there, with your adversary draped over you and bleeding on your business suit. You cannot move back because of the terrain, but you can move to the side. Move either to the left or right, as the situation requires—but move. When you obtain the necessary standoff distance to the side, you may then extend your firing position to the ready. The disadvantage of the speed rock is that it unbalances you to the rear, but this is not as crucial as many critics make out. Keep the proper perspective on the matter: you are close enough to your enemy to

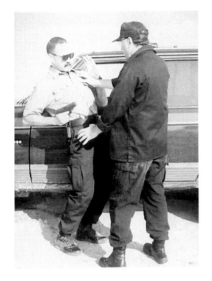

The only solution is the Taylor-designed speed rock, which gets the pistol on target quickly, as well as preventing the adversary access to your weapon.

The second distance interval is within arm's reach of the adversary, but still allows some room to maneuver.

smell the onions he had for lunch, you *cannot* move back, and you must shut him down immediately or you die. Only the speed rock will let you do that. The advantages are that it is lightning quick (two shots in less than one second), and it denies an adversary access to your pistol. It is a special-purpose technique for special circumstances. If the choice is to unbalance yourself and shoot your opponent in the nick of time or to be too slow and die, which would you choose?

Notice that extending your weapon into a standard position places it right into the adversary's hands.

Similarly, attempting to gain any standoff distance before firing is often a futile move when the adversary is so close, since he will simply walk forward and fill the void you've created.

When you are within arm's reach but have ample room to maneuver to the rear for standoff distance, the speed rock is not the preferred technique. In this interval, the close-contact technique developed by the Los Angeles Police Department is a more suitable choice because it allows you to maintain balance while you also protect your firearm from an adversary. However, the close-contact technique is slightly slower than the speed rock. Notice also that in situations where the speed rock can be employed, the close-contact technique is less desirable because of both speed and weapon retention issues.

After the adversary has been rendered safe with the close-contact technique, you should put distance between you and your adversary in the event of a failure to stop. You do this by executing the step-back technique. However, one step back is generally not enough because your adversary will simply step forward and fill the void that you created. And if you extend the pistol in a standard firing position, you have just placed the

The only tactically sound solution is to use the Los Angeles Police Department's close-contact position.

After the immediate problem has been neutralized, stepping back to gain distance is a good idea.

muzzle in his very grasp. If you have the room, step to the rear with the firing-side leg as the firing hand finds the pistol and then step back again with the support-side leg, reestablishing the firing platform, as you present your pistol and fire. A second series of steps after the shots have been fired and during the target assessment phase will create more distance in the event of a failure to stop.

The third critical distance is when the adversary is just outside of arm's reach— or one step away from contact.

As the fight begins, your opponent must take a step forward to press the assault, so you can use the distance to step back, creating enough distance to . . .

. . . be able to shoot from a conventional position.

After the threat has been neutralized, lower the pistol out of your line of sight to determine if more gunfire is needed.

It is a good idea to include a second set of steps to the side or rear to gain even more distance. You'll need it if a failure to stop occurs.

The classic speed rock scenario: close proximity to the adversary and no room to move back at all!

After the execution of the speed rock, don't just stand there! Begin to create distance in the only direction available, probably to the side.

Then take another step, just in case the fight is not over. Notice that the pistol is immediately placed in a two-handed Weaver ready position as soon as stand-off distance is achieved. Notice also that the weapon is kept oriented on the threat as the operator moves clear.

The interval of the confrontation and your ability to extend that interval will determine your response to an assault. When the distance between you and your adversary extends beyond his arm's reach, the speed rock or close-contact position is generally ineffective, and the step-back technique is usually recommended. However, the solution you select will be based on your perception of this interval. Theorists who've never faced death or seen their opponents' muzzle flash often attempt to discount close-quarters defenses. They seem to forget that each particular problem has its own specific solution. If you want to be successful (i.e., keep breathing), you must make your solution fit the problem. Don't make the mistake of trying to force one technique to fit every circumstance, because you will fail.

If you need to drive a nail, get a hammer; if you need to turn a nut, get a wrench. Similarly, if you need a speed rock, don't try to make a step back fit the problem—or worse yet,

just ignore the necessity of a close-quarters defense. The price
of failure in close combat is too high.

FIFTEEN

OUTDOOR TACTICS AND MOVING THROUGH OPEN AREAS

—————◆·◆·◆—————

I heard the bullets whistle: and believe me, there is something charm-
ing in the sound.

George Washington
Letter to his mother, 3 May 1754

So far we've discussed situations involving indoor confrontations. Only in extremely unusual circumstances will an urban busi-
nessman or homeowner have reason to pursue an adversary outside his "castle" once the adversary has fled. Neither will there be
any urgency to go check and clear yards or external perimeters. If
you are inside and your adversary is outside, your best solution is
to call the police and barricade yourself behind cover.

The story changes, however, for the police officer. A police
officer has to go and search for these people. Similarly, the rural
citizen is not likely to call out the troops for every little suspicious sound in the back 40. There are times when you must go
and hunt under the open sky and check things out for yourself.
This is not a study of military squad tactics. Although some concepts have come from that discipline, the tactics described here
are designed for the lone operator and the small two- or three-
man team.

Tactics for moving through and searching outdoor open
areas are very similar to those used indoors. The only real differ-
ences are that the distances involved are longer and the spaces
are bigger. Additionally, the problem is not contained within
four walls. Subsequently, outdoor problems are somewhat more
difficult to solve because there are more potential danger areas
than there are inside a structure.

Outdoor, long-range activities may require more flexible weapons than those favored for indoor scenarios.

The low ready position, demonstrated here by four-weapon combat master Don Busse, is favored for a balance of visibility and reaction speed.

Submachine gun students learn long-range engagement techniques with a weapon intended for close quarters.

Using cover *as* concealment or simply using concealment is mandatory. You might be able to observe an entire area from your hidden position of cover. Remember, you are looking for target indicators. Initially, look for places where it is likely that an adversary might be hiding. You will be moving toward those areas to clear them. Before you move from your covered position, select the covered position you will be moving to next. Such a move must be designed to gain you some type of an advantage. Seek either an enhanced zone of fire or a better look into the area you are checking. Do not abandon your position out of impatience.

Visually scan along a varying vertical axis or from close to far and back. This will allow you to examine the same spot from more than a single angle. Such a visual search will enhance the possibility of seeing something you missed during previous passes of the same area. Look slowly and carefully; don't just pass your eyes over the area haphazardly.

Avoid silhouetting yourself or giving off a shadow. Be aware of the light source and stay out of it. This may not be possible,

When operating outdoors, stay below the visual horizon and do not "skyline" yourself on rooftops or ridgelines. Additionally, avoid backgrounds that contrast sharply with the clothing worn. (Photo courtesy of Chuck Taylor.)

but you must try nevertheless. If there is a light illuminating the area, stay below it and out of the light wash. During the day, stay off fences and rooflines that will place the sky or other light surface behind you. Additionally, do not pause in front of light-colored walls. Such things make your outline stand out like, well, a target.

Be careful about such reflective outside surfaces as windows. Such surfaces may be useful in locating an adversary, but they may just as easily be used by him to locate you.

Keep low when you move from point to point. You should be able to shoot on the move as soon as you leave cover. Stealth is still your greatest asset. Don't run unless you are already under fire. It is very difficult to assess potential danger areas while running. Also, if you decide to change directions or if you must shoot while moving, you cannot readily do either one at a dead run.

Be alert to any channelized areas, such as alleys, driveways, and spaces between buildings. Such areas are perfect for ambushes because there is only one way in or out. Once you commit yourself to such an area, there is no easy way to abandon it.

When moving with a partner (or two) you may use one of two methods borrowed from squad tactics. The first one is

Realize that your adversary may also know about camouflage and concealment and try to blend into the background. (Photo courtesy of Chuck Taylor.)

If you know where your opponent is but he does not know where you are, you can launch your own attack from concealment.

Outdoor operations may require a greater variety of tactics and shooting positions than the indoor environment. Don Busse instructs students in the prone position and the close-quarters/underarm assault position—two extremes in engagement technique.

termed "traveling overwatch." This method has the team members moving to static positions as they cover divergent areas of responsibility. This is the same method used indoors as the team moves from point to point. The advantages of this method are that it places two or more guns and eyes on the danger area instead of one. Additionally, all the team members are in a position of mutual support, and all can fire at a sudden threat without exposing themselves to their own team members' gunfire.

The second method is called "bounding overwatch." This may be used to enter or exit a very hazardous environment where there has already been some shooting or where a gunfight is expected. Additionally, this technique is extremely useful when evacuating an area (particularly if you are evacuating a downed team member).

Bounding overwatch works this way: one member of the team moves to a position of cover, his partner then leapfrogs him to another different point of cover, and then the next man will move past his partner to a farther point and cover his partner's approach. This can be done at any speed, depending on the circumstances. The distance each partner travels is dictated by the terrain itself, but it should not be more than 15 meters. Once you are committed to moving, do not stop. If the original plan of movement becomes untenable, you can always go back. But simply stopping in the open is foolish.

Chuck Taylor demonstrates the high ready carry position during a submachine gun course.

A student moves through wooded terrain with a Heckler & Koch MP5.

The ultimate in outdoor/long-range coverage! (Photo courtesy of Chuck Taylor.)

Some circumstances may arise suddenly, so learn to deploy your resources immediately. This shooter is learning counterambush techniques. (Photo courtesy of Chuck Taylor.)

Outdoor problems often require supported shooting positions. Learn these positions with whatever weapon system you use.

SIXTEEN

TOOLS FOR SEARCHING

And he that hath no sword, let him sell his garment, and buy one.

Luke 22:36

We've all heard the story of the idiot who brought a knife to a gunfight, but there are also plenty of stories out there about people who similarly went into battle without the proper equipment. This is even more poignant when we realize that the proper gear was at hand and the people in question had prior warning of its probable impending need. We can only speculate, for example, what a pair of Gatling guns might have done for Custer on one fateful day in 1876.

A tactical search of any kind is a proactive event. This means that it generally doesn't just happen unexpectedly, like a surprise gunfight. Rather, you go in search of the fight . . . on purpose. You generally know what you are getting into. Therefore, you want to bring the right gear with you to the party.

Some weapons, for example, may be more suitable for certain types of searches. If you are a rural ranch owner about to search the area around your home for a possible trespasser, the AR-15 or SKS by the back door will do just fine. In fact, such weapons are preferred for those environments because of their greater reach, power, and versatility. A carbine or compact repeating rifle is very useful if you expect contact at greater than 25 meters. I know some Korean businessmen who were glad to have had their "ugly and sinister" military-style rifles a few seasons ago in "kinder, gentler" L.A.

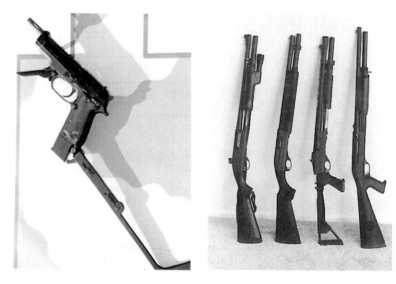

Remember what it is that you are trying to accomplish. Some weapons are neat, but this does not mean that they are the best tool for the job.

The inner-city apartment dweller is in a different position altogether. He may want to limit himself to an easy-to-use shotgun with low-penetration birdshot. His apartment may not be large enough to even bother searching since he might be able to survey the entire area from his bedroom door. He might choose a weapon that he or his unskilled family can hide behind until the police arrive. Additionally, he may have concerns over rounds that miss, disrupting his neighbors' living quarters, so single-projectile weapons such as pistols may not be a good choice. Carbines and rifles would, likewise, not be a consideration for this application.

At other times, the ubiquitous service pistol will be the tool best suited for the job since it is a weapon that most people already have access to and can readily practice with at urban indoor ranges. The pistol's maneuverability and compactness are also the reason that many people, who have access to other weapons as well, choose it when they must operate in close-quarters environments. For example, if I am about to search a typical urban structure I prefer the pistol. I know, I know—the shotgun, the carbine, and the submachine gun are all more power-ful, reach longer, etc., but they are also very l-o-n-g weapons. The pistol is much handier and more maneuverable in the indoor environment than the other weapons available. For instance, you can easily control it with one hand, if necessary, and you can move with it through areas that would severely limit your mobil-ity if you were armed with a long gun.

If I am searching an outdoor area, on the other hand (such as a backyard, an alley, or a dark beach), I'll take a shotgun or a submachine gun. You will not face the same closely confined movement outside that you will in the indoors environment, so the length of the shotgun or submachine gun will not create a liability. The advantages in stopping power as well as in speed must not be ignored for rapidly moving outdoor targets. The point is that you must first examine the "battlefield" you will most likely be operating in and then select the tools to fit the problem at hand.

There is another item that is mandatory for the searcher: a compact, high-intensity flashlight. Every policeman should have one on his belt; every homeowner should have one by his home defense weapon. If your gun is to be used exclusively for defend-ing your home or business, you can get away with a dedicated weapon mount as designed for special operations/SWAT use.

If your gun will see double duty as a concealed piece that you stuff into your waistband every morning, either avoid the weapon-mounted light or, better yet, get a second identical pis-tol. If you only have one gun, secure one of the compact Laser Products Sure-Fire or Streamlight Stinger flashlights and devel-op a strong flashlight-ready position.

Study your environment and the realities that you might face. Develop a plan of action, train your partners and your fam-ily, and be ready. Make sure that the next urban terrorist who hides from you will be in more danger from you than you are from him.

A standard submachine gun is often one of the best and most versatile tools available for indoor and limited-range outdoor scenarios.

Even with the modern advances in weaponcraft, the ubiquitous handgun is still often preferred for indoor scenarios because of its controllability and maneuverability.

Dale Fricke oversees a firing drill with a suppressed submachine gun.

SEVENTEEN

HOW TO
PRACTICE TACTICS

———◆◆◆———

*Practice doesn't make perfect; only PERFECT practice makes perfect.
If you practice crap for twenty years, all you'll be is a "crapmaster."
Crapmasters do not win gunfights!*

Marc Fleischmann

By now, most of you know that one of the most important things you can do to enhance you firearms skills is constant and consistent dry practice of the fundamentals in conjunction with regular live-fire practice. Those who practice the "three secrets" as well as gun manipulation drills (e.g., presentation, reloading, malfunction clearances) on a regular basis fare much better in combat than those who do not. Historical evidence and personal experience will prove that fact quite easily. The study of personal tactics, particularly as they relate to building searches, may be handled in the same fashion.

At this point in your studies, you should know how to handle most of the architectural obstacles you may face in a typical urban building. If you wish to make these skills second nature—instead of simply having knowledge about their execution—you must "dry practice" them. In every apartment or home you will have doors, corners, hallways, hallway intersections, and (if you are lucky) stairs.

First of all, follow the standard dry practice procedures to unload your firearm (and secure the ammunition). The last thing you want is any new ventilation at the end of your dry practice session. Not only is this embarrassing, but it also annoys the neighbors.

Training in a tactical simulator is essential to a thorough understanding of tactical principles.

With every step, the trainee is presented with another potential danger area, threat, and shoot/no-shoot decision.

Next, find a door. Practice approaching the door, checking the doorknob, and opening the door while maintaining a ready position. Practice clearing the room beyond from the outside and getting through the door (do the latter several times with the door opening inward as well as outward).

Now, find a corner and repeat the process. Clear a right-hand corner 10 times and then reverse it and clear a left-hand corner. Watch your movement on the approach and be careful about not allowing your weapon to precede you into unsecured space.

Find a T-intersection and repeat the process again. Pay attention to the basics. Read the discussion on T-intersections again to be sure you know how to handle them properly. Just as you do with simple dry practice, proceed slowly and make the movements perfect. This quest for perfection will yield extreme smoothness, which will also make you efficient. Remember, there is little need for *speed* in most tactical situations.

When you are comfortable with these techniques, introduce a dry practice target into the equation. This target should be a depiction of a hostile human being, not simply a neutral silhouette. This will accustom and condition your mind to search for humanoid features.

The next step involves the assistance of a partner. For many years my wife and I were on different working schedules; I would arrive home about an hour after she'd left for work. Before she left, however, she would usually hide a handful of evil-looking humanoid targets within the house. When I arrived home after a long night at "The Front," I would unload my pistol and follow my dry practice procedures, and then I would hunt the hostiles. She would make them extremely difficult for me to find, because if I failed to locate a target I had to take her out for an expensive dinner and night on the town. Of course, I was required to be honest about whether I ever failed to find one.

Some fainthearted readers may be put off by such extreme measures and practices. Well, all I can say is that when your life depends on your skills, nothing is extreme!

The next level is to play a tactical hide-and-seek game with plastic "red guns." We conduct similar exercises at our training courses, such as the Advanced Tactics Course. Have your part-

Simulations may be executed by the individual or team.

Above all, it is important to learn how to move through each obstacle smoothly and correctly.

ner go hide in your residence (armed with a plastic red gun prop) as an adversary would. You will go search for him as you would if you were really hunting a dangerous adversary. **NOTE:** Please use only nonfiring plastic training guns for this exercise!

For those who are fortunate enough to have access to Simunition FX training cartridges, you may set up actual force-against-force scenarios. These Simunition FX cartridges may be fired in real firearms if you first install a safety kit. This type of training is as close as you can get to an actual building search for an adversary or an actual gunfight. One caution about Simunition: the strike of these cartridges stings considerably. The tendency is to armor yourself from head to toe in hopes of mitigating the sting; I am a firm believer in the concept that pain is the greatest of teachers. Other than protecting the obvious vital areas (eyes, throat, groin), a simple long-sleeved shirt and pants will suffice.

Additionally, the great motivator here is the possibility of being hit and the resulting pain. If you use Simunition too

Finally, you may graduate to a tactical "hide and seek" where your training partner prepares an ambush while you go find him.

much, you will become desensitized to the sting, and it will lose its training value. I've seen this firsthand in SWAT training where officers will expose themselves to the incoming fire because subconsciously they are no longer afraid to get hit. This has disastrous potential on the street.

Just as with any form of semiathletic skill, continuous practice is important to the development of tactical skills. But not just any type of practice will do. Only perfect practice makes perfect performance. This applies to both the physical and the mental-emotional aspects of tactics. Perfection—unattainable as it is—must be your goal when your life depends on your success.

I remember well the thought processes I experienced during my first gunfight (my first real brush with death) years ago. In spite of the very real life-threatening danger, the smoke, the gunfire, and the confusion, I reacted tactically, mentally, and emotionally just as I had trained myself to do in tactical simulations. When I heard my adversary's gunfire I remember thinking, *This is familiar territory . . . I've been here before . . . I know what is going to happen next.* Such practice as I've described will develop "experience" as much as is possible in today's world without really "getting in the elephant's face." The experience obtained in training can be as *real* as you wish to make it. It is simply a matter of attitude. Often, that very experience, obtained on the training field, is what will define the difference between a glorious victory over incredible odds or an unexpected death on the battlefield.

TERMINOLOGY OF TACTICS

——◆◆◆——

Ace: A master tactician who has developed an extremely high level of skill and expertise in tactics and weaponcraft.

Angle: The angle of visibility and fire obtained as an operator moves through a structure.

Angular Search: An inspection around an obstruction, such as a corner, focusing on the apex of the obstruction and then proceeding incrementally until the entire area beyond it is visible.

Area of Responsibility: The physical sector of coverage for a member of a tactical unit. Each area of responsibility may contain one or more danger areas. Areas of responsibility coincide with sectors of fire.

Breaking Planes: Geometric planes are encountered when reaching a threshold into unsecured areas (space). Breaking into a plane is the act of penetrating that unsecured space.

Combat Countermeasures: Alternative force techniques used by an armed operator to regain physical control of the situation when shooting is contraindicated.

Concealment: Anything that an operator can use to hide himself. Concealment is not bullet resistant.

Condition Orange: The mental state of specific alert.

Condition Red: The mental state of combat response, awaiting the mental trigger.

Condition White: The mental state of inattention.

Condition Yellow: The mental state of relaxed alertness.

Cover: Anything bullet resistant that an operator can hide behind to protect him from hostile gunfire. Cover may be used as concealment.

Danger Area: A location or position that is likely to conceal or contain a hostile.

Dividing Attention: The necessary division of focus by a lone operator between two danger areas located at opposite angles.

Extreme Angle: The unseen angle found along the walls on each side of a door as seen during an angular search from the outside.

Fatal Funnel: The area directly in front of a portal, such as a door, and also extending back from it for a few feet. Operators must avoid lingering in this area because it allows a hostile in the danger area to shoot them at will.

Kill Zone: A circumscribed area under the control of hostile gunfire and limited by the skill of the adversary, the range of his weapon, and natural or man-made obstructions.

Mental Trigger: Any action or condition exhibited by the hostile and predetermined in a shooter's mind that will cause a reflexive lethal response.

Mr. Murphy: The mythical nemesis of all tacticians. Murphy is an imaginary being who is tasked with making sure that anything that can possibly go wrong, does . . . at the worst times imaginable.

Soft Check: A surreptitious check of a doorknob to ascertain its status (locked or unlocked) before getting committed to opening the door.

Soup Sandwich: A tactically incompetent individual (e.g., clam chowder on sourdough).

Tactics: The art of maneuvering skillfully against an opponent and toward a desired objective.

Target Indicators: A term taken from the sniper discipline describing anything that may indicate the presence of a hostile inside a search area.

Theorist: One who thinks he knows about practical, hands-on tactics, but in reality knows nothing but untested classroom theories.

Three-Eye Principle: The concept of maintaining the position of your weapon's muzzle in accordance with the direction of your visual scan. This principle enhances your reactions to suddenly appearing threats.

Three Secrets: A term coined by Chuck Taylor that indicates the three imperative factors necessary for hitting your adversary under stress before he hits you: sight alignment, sight picture, and trigger control.

Vertical Method: The scanning method preferred for visually searching a danger area. It involves visually scanning along a vertical axis—in and out instead of side to side.

SOURCES FOR GEAR
AND TRAINING

Laser Products
18300 Mt. Baldy Circle
Fountain Valley, CA 92708-6117
Sure-Fire tactical flashlights and flashlight mounts for weapons. If you need a light, you can do no better than the Sure-Fire series.

M-D Labs
8333 Pecos Dr. #4
Prescott Valley, AZ 86314
World-class holsters and other tactical gear formed out of Kydex by folks who know what the armed professional really needs.

Trijicon Sights
P.O. Box 6029
Wixom, MI 48393-6029
Radioactive night sights for all weapon types.

Internet Sales
Second Amendment Mall—http://www.2ndmall.com

The author is available for consulting, weapons and tactics training, or for speaking and lecturing engagements. He can be contacted in care of Paladin Press, P.O. Box 1307, Boulder, CO 80306.

SUGGESTED READING

Morrison, Gregory. *The Modern Technique of the Pistol.* Pauldin, Ariz.: Gunsite Press, 1991.

Musashi, Miyamoto. *A Book of Five Rings.* Translated by Victor Harris. New York: The Overlook Press, 1974.

Suarez, Gabriel. *The Tactical Pistol: Advanced Gun Fighting Concepts and Tactics.* Boulder, Colo.: Paladin Press, 1995.

_____. *The Tactical Shotgun: The Best Techniques and Tactics for Employing the Shotgun in Personal Combat.* Boulder, Colo.: Paladin Press, 1996.

Sun Tzu. *The Art of War.* Translated by Samuel B. Griffith. London: Oxford University Press, 1963.

Taylor, Chuck. *The Combat Shotgun and Submachine Gun.* Boulder, Colo.: Paladin Press, 1985.

_____. *The Complete Book of Combat Handgunning.* El Dorado, Ark.: Desert Publications, 1982.

_____. *The Fighting Rifle.* El Dorado, Ark.: Desert Publications, 1983.

_____. *The Gun Digest Book of Combat Handgunnery.* Northbrook, Ill.: DBI Books, 1997.